Human Systems Interactions

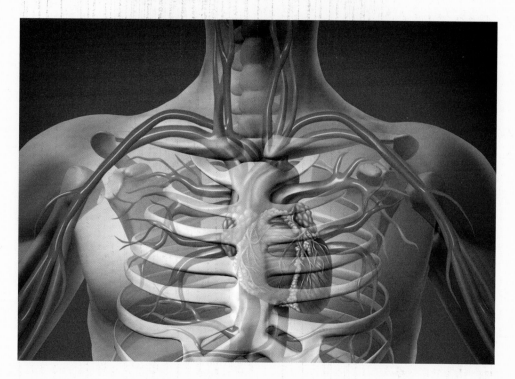

Developed at
The Lawrence Hall of Science,
University of California, Berkeley
Published and distributed by
Delta Education,
a member of the School Specialty Family

1465674
978-1-62571-179-3
Printing 7 —12/2021
Sheridan Wisconsin, Madison, WI

Table of Contents

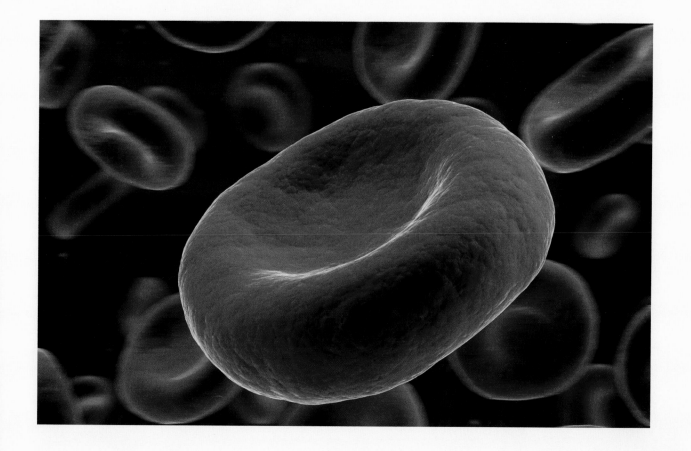

Human Organ Systems

The sections in this article introduce one of the truly exceptional systems in the world. That system is your body.

Your body is essentially the same as all of the other humans in the world. Every body is a system composed of the same fundamental organ subsystems. Most of them will be somewhat familiar to you. They are the **circulatory system**, **digestive system**, **endocrine system**, **excretory system**, **muscular system**, **nervous system**, **respiratory system**, and **skeletal system**.

You are about to begin an exploratory experience into the structure and function of human organ systems. You will become a specialist in one of these systems. You will develop knowledge about how one system functions and how it interacts with each other system. And you will become aware of **symptoms** that signal a problem with the system. This information will help you work with other system specialists to diagnose a problem.

Complex, interacting organ systems work together to make every human action and activity possible, from reading this page to taking a heart-pounding run on the beach.

Circulatory System

As your **heart** pumps blood through the body, you can feel the blood vessels pulsing just under your skin. Feel for the pulse in your neck or wrist. Your pulse shows how fast your heart beats.

The circulatory system consists of the heart, blood vessels, and blood. This system, also called the cardiovascular system, transports water, nutrients, **hormones**, and oxygen to the **cells** in your body. It carries waste products such as carbon dioxide and water away from your cells. It also helps protect your body from infection and disease.

The circulatory system can be described as a larger system with two subsystems: the cardiovascular system and the lymphatic system. This discussion is only about the cardiovascular system.

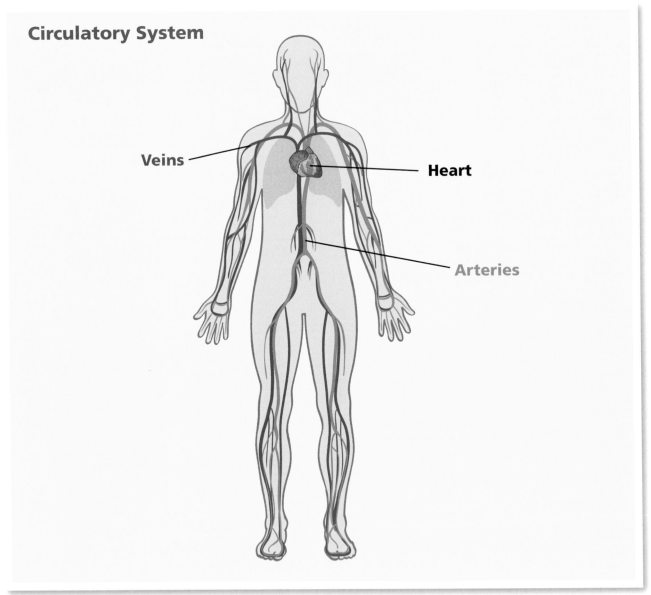

Circulatory System

Veins

Heart

Arteries

The circulatory system includes the heart, blood, and blood vessels. The arteries deliver oxygen, water, and nutrients to the body's cells and the veins carry away waste materials.

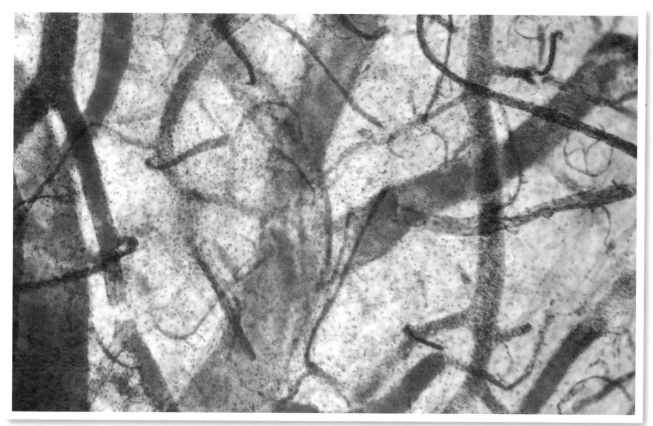

Arteries, veins, and capillaries form the body's 100,000-kilometer-long network of blood vessels. This photograph shows blood vessels as seen under a microscope.

Parts of the Circulatory System

The heart. The heart is slightly left of the middle of your chest. It is protected by the sternum and rib cage. The heart is about the size of your fist. It is made up of strong **cardiac muscle** tissue. It pumps about 5 liters (L) of blood throughout your circulatory system every minute.

Blood vessels. Blood vessels carry blood from your heart to the cells in your body and back again. The three major types of blood vessels are **arteries**, **capillaries**, and **veins**. Arteries carry blood away from the heart. They have thick, muscular walls. Capillaries are only one cell wide and have very thin walls. They exchange nutrients, gases, and waste products with cells. Veins return blood to the heart. The blood vessels all connect in a giant network.

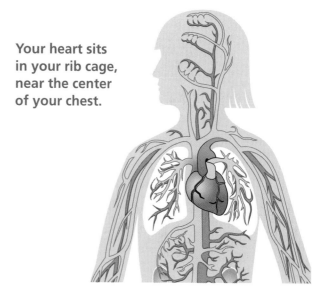

Your heart sits in your rib cage, near the center of your chest.

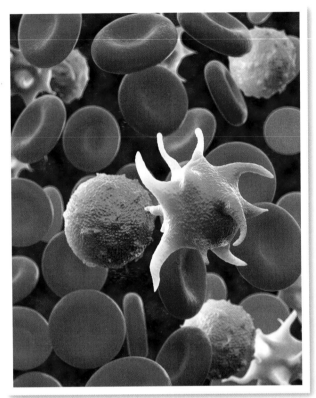

White blood cells (white, round) help fight infections, while red blood cells transport oxygen. Platelets (white with spikes) help to clot blood.

Blood. Blood contains three types of blood cells: **red blood cells**, **white blood cells**, and **platelets**. All three types are produced in **bone marrow** of long bones in the legs and arms. These cells are suspended in a fluid called **plasma**. Blood plasma is mostly water. Plasma and the blood cells transport substances such as sugars, hormones, and gases.

The most abundant cells in your blood are red blood cells. They contain an iron-rich protein called hemoglobin, which "captures" and transports oxygen and carbon dioxide. White blood cells are part of the immune system. They help defend the body against disease. Platelets are important for blood clotting.

Functions and Interactions

The circulatory system interacts with every other organ system. It transports nutrient molecules from the digestive system, oxygen from the respiratory system, and hormones from the endocrine system to virtually every cell of the body. The nervous system controls the pumping of your heart.

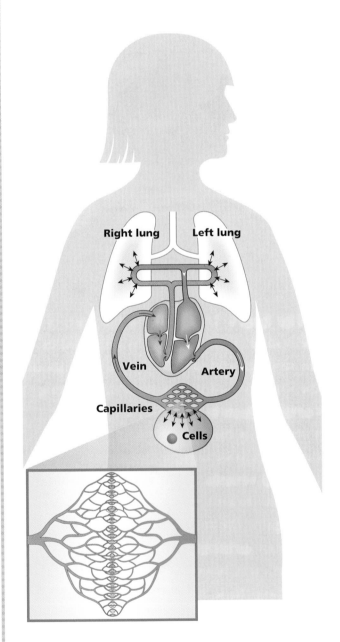

Capillaries form netlike structures throughout the body's tissues to reach cells.

Transportation. The heart pumps blood to the cells in the body. On its way, blood picks up oxygen in the lungs, and nutrient molecules from the small intestine. Arteries carry oxygen and nutrient-rich blood. In capillaries, oxygen and nutrients transfer into cells. Products such as carbon dioxide and excess water are transferred from the cells into the capillaries.

On the return trip to the heart, blood travels in veins. The blood carries waste products from the cells to the kidneys of the excretory system. The kidneys filter out wastes. The blood continues to the heart. It is pumped a short distance to the lungs, where it exchanges carbon dioxide for oxygen.

The endocrine system depends upon the circulatory system to transport hormones throughout the body. When you are startled, for example, adrenal glands release a hormone called adrenaline into the bloodstream. The blood carries the adrenaline to the heart. The hormone causes the heart to beat faster. Adrenaline also increases blood pressure. High pressure increases the delivery of nutrients and oxygen. As a result, the body produces more energy.

Strenuous exercise involves much more than just muscles. The digestive, circulatory, respiratory, nervous, skeletal, and other organ systems are also at work.

Temperature regulation. The circulatory system helps regulate body temperature. When you are cold, blood circulation slows near the skin surface. This conserves core heat. When you are hot, blood flow to the skin increases. The body transfers thermal energy into the environment. This temperature response can make a cold person look pale and an overheated person look red. The blood can also release extra water as sweat. Sweat cools your body through evaporation.

Protection from disease. White blood cells help your body fight infections. Each of the five kinds of white blood cells has a purpose. Some fight bacteria, fungi, viruses, and parasites. Others are part of immunity and allergic responses. White blood cells can leave the circulatory system and go directly into the tissue where the infection is. If you have ever seen pus from a healing wound, you have seen the remains of white blood cells involved in the fight against infection.

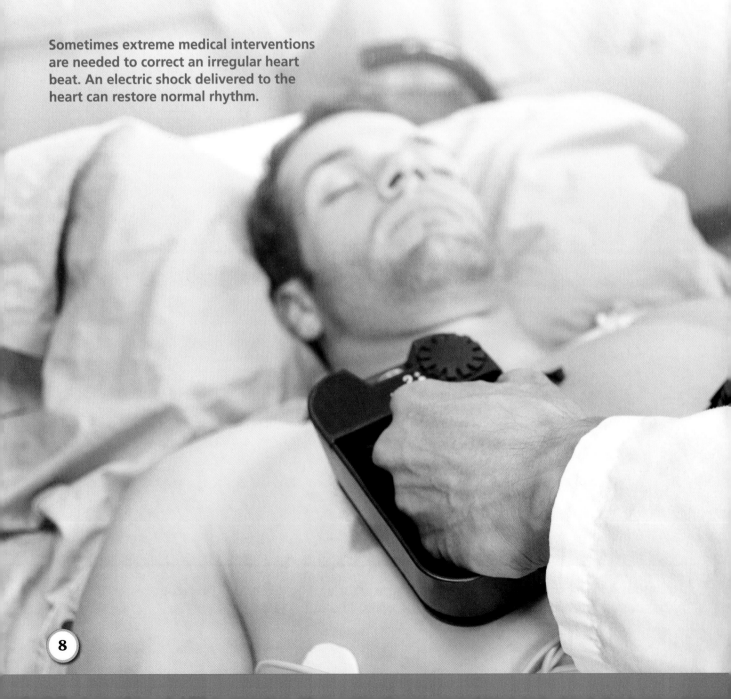

Sometimes extreme medical interventions are needed to correct an irregular heart beat. An electric shock delivered to the heart can restore normal rhythm.

Service to important organs. The brain and the heart get top priority for blood supply. Without blood flow, heart muscles stop working. This can cause a heart attack. Lack of blood flow in an area of the brain can cause a stroke. Strokes damage brain tissue, temporarily or permanently.

Anatomy of a Heart Attack

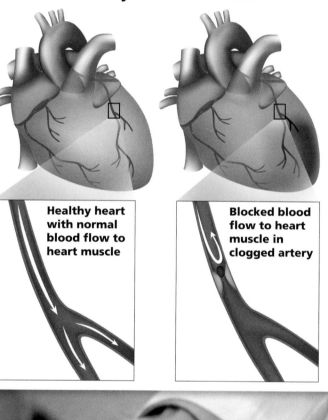

Healthy heart with normal blood flow to heart muscle

Blocked blood flow to heart muscle in clogged artery

When Things Go Wrong

These symptoms can mean that something is wrong with the circulatory system.

- **Fatigue**
- Headache
- Chest **pain**
- **Abnormal** blood pressure
- Dizziness and fainting
- Poor healing of wounds
- Abnormal pulse
- Cold or numb hands and feet

With prompt medical treatment, many people recover fully from a heart attack.

Digestive System

Why does your stomach growl? Can you make it stop growling? Your stomach is part of the digestive system. This system usually carries on without our noticing it. The tubes and organs of the digestive system transport food. The digestive system breaks food into nutrients and removes wastes.

Parts of the Digestive System

Consider the digestive system as one long tube. Food enters the mouth, travels down the esophagus, enters the muscular stomach, and moves into the small intestine. The small intestine is about three and a half times the length of your body. Most of the nutrient absorption happens here. Whatever is left moves into the large intestine, which wraps around the small intestine. Finally, waste passes into the rectum and leaves the body through the anus. Other organs, especially the liver, and **glands** contribute to the digestive process.

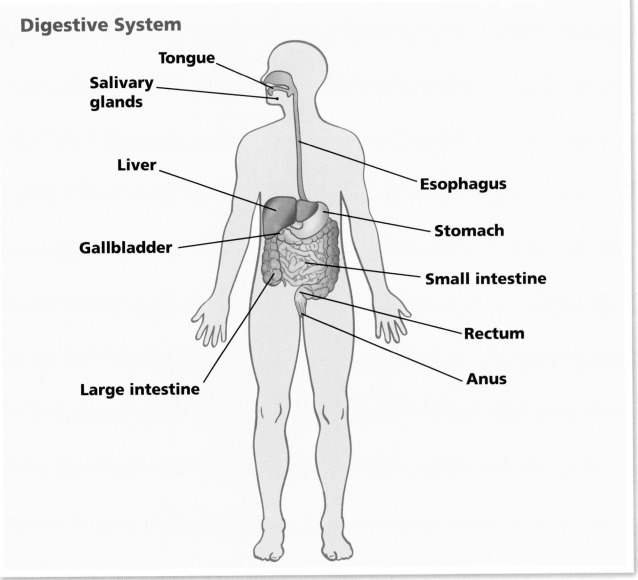

Digestive System

Tongue
Salivary glands
Liver
Gallbladder
Large intestine
Esophagus
Stomach
Small intestine
Rectum
Anus

In the digestive system, food is broken down into nutrients the body can use to support growth or to release energy.

Functions and Interactions

Food provides energy for our bodies. It contains the building blocks (carbohydrates, proteins, and fats) that our body needs to grow and to repair damage. Our body can use food only after the food breaks down into nutrient molecules.

Food breakdown. Your mouth breaks food down physically and chemically. Teeth, jawbones, jaw muscles, and your tongue mash up your food. In the mouth, salivary glands produce **saliva**. Saliva contains **enzymes** that start chemically breaking down the food. The **smooth muscle** tissue of the esophagus contracts in a wave-like action called **peristalsis**, pushing the food into the stomach.

A healthy, balanced diet includes grains, fruits and vegetables, dairy, protein foods, and fats. What's on your plate?

The muscular stomach mixes the food with enzymes and acid. As the liquid moves into the small intestine, enzymes from the pancreas are added. (The pancreas is part of the endocrine system.) The gallbladder supplies bile. Bile contains acids that help dissolve fats. After a couple of hours, the food has turned into a thick liquid. It slowly enters the small intestine, moved along by more smooth muscle.

Most of these muscle contractions and the release of digestive enzymes are controlled by the autonomic nervous system. They happen when you are asleep or even if you are in a coma.

And that stomach growling? Muscle contractions churn and break down food in your stomach. When you swallow food, you also swallow air. The growl is the sound of the air getting squeezed in your stomach.

Fruits are sources of many essential nutrients, such as potassium, vitamin C, fiber, and folic acid.

The foods you eat must be broken down by mechanical and chemical processes in your body to make nutrients accessible to your cells.

Absorption of nutrients. The small intestine breaks down food even further, into nutrient molecules. The small intestine is the longest part of your digestive tract, about 7 meters (m) long. It is wrapped with blood vessels of the circulatory system. The blood vessels absorb the nutrient molecules and carry them to the liver. The liver processes the nutrients, stores them, and releases them into the blood to be carried to the **cells** in the body. The liver also breaks down toxic chemicals, such as alcohol, before releasing them to the excretory system.

Hormones from the endocrine system determine the speed at which food moves through the digestive system. Slower digestion means more complete absorption of nutrients. Hormones also regulate how much sugar enters your bloodstream. And they tell your brain when you are full.

The large intestine houses many kinds of bacteria. These produce vitamins that are essential for the functioning of the nervous system and other body processes.

Excretion of wastes. After nutrient molecules are absorbed into the blood from the small intestine, whatever food matter is left enters the large intestine. There, water passes into the bloodstream. Solid waste is stored in the end of the large intestine as feces. A signal to the brain to eliminate the solid waste occurs when the waste moves into the rectum. The digestive and excretory systems work together to balance the water in the body. If you are dehydrated, the large intestine transfers more water from the digested food to the blood. As a result, your waste products get more concentrated.

When Things Go Wrong

These symptoms can mean that something is wrong with the digestive system.

- Abdominal **pain**
- Nausea
- Vomiting
- Diarrhea or constipation
- Burning in the throat or chest
- **Abnormal** swelling or bloating

Mouth

Esophagus

Right lung

Left lung

Heart

Cells

Stomach

Small intestine

Large intestine

Colon

Rectum

Anus

The digestive system rids the body of solid waste and helps maintain water balance.

Your body loses water all the time. You can stay hydrated by drinking plenty of water.

Endocrine System

Have you heard people complain that teenagers' **hormones** are out of control? Well, don't worry. Teenage hormones are doing just what they are supposed to do. Hormones are part of the endocrine system.

The endocrine system influences your entire body: every **cell** and every organ. It is responsible for reproduction, growth and sexual development, energy level, and response to stress and injury.

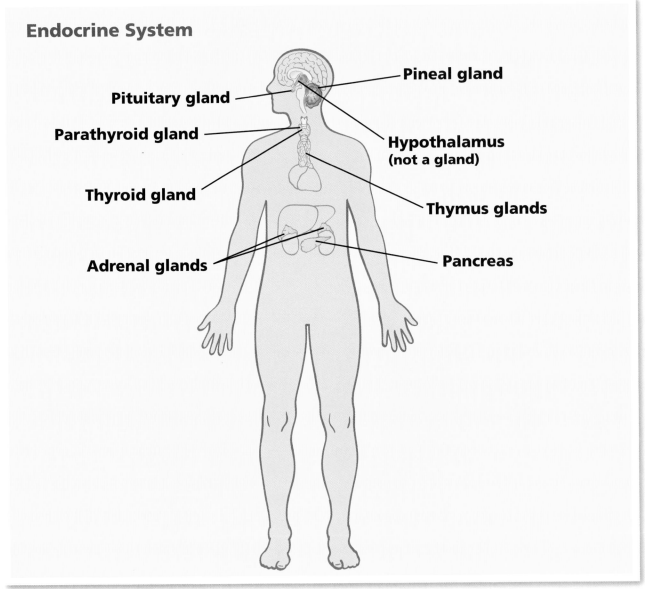

Endocrine System

- Pineal gland
- Pituitary gland
- Parathyroid gland
- Hypothalamus (not a gland)
- Thyroid gland
- Thymus glands
- Adrenal glands
- Pancreas

The glands that make up the endocrine system produce and release hormones that control many important body functions.

Parts of the Endocrine System

Glands. The specialized organs that manufacture and release chemicals are called **glands**. Some glands, such as the pancreas, release enzymes into the digestive system and hormones into the blood. Other glands primarily release hormones into the blood, which carries them through the body. The pituitary gland affects many organs and processes. The pineal gland is still a mystery, but seems to have a role in sleep. The thyroid, adrenal, and parathyroid glands regulate **metabolism** and minerals. The adrenal glands also are involved in stress response.

The thymus gland is involved in immunity. The reproductive glands (ovaries in women and testes in men) are the main source of sex hormones.

The hypothalamus is attached to the pituitary gland in the brain. It links the nervous and endocrine systems and makes hormones that control the pituitary gland.

Hormones. Hormones are like a messaging system that delivers instructions via the bloodstream. They are chemicals released in one part of the body that travel to another part of the body to stimulate action.

The endocrine system maintains your metabolism and releases hormones that help you digest food.

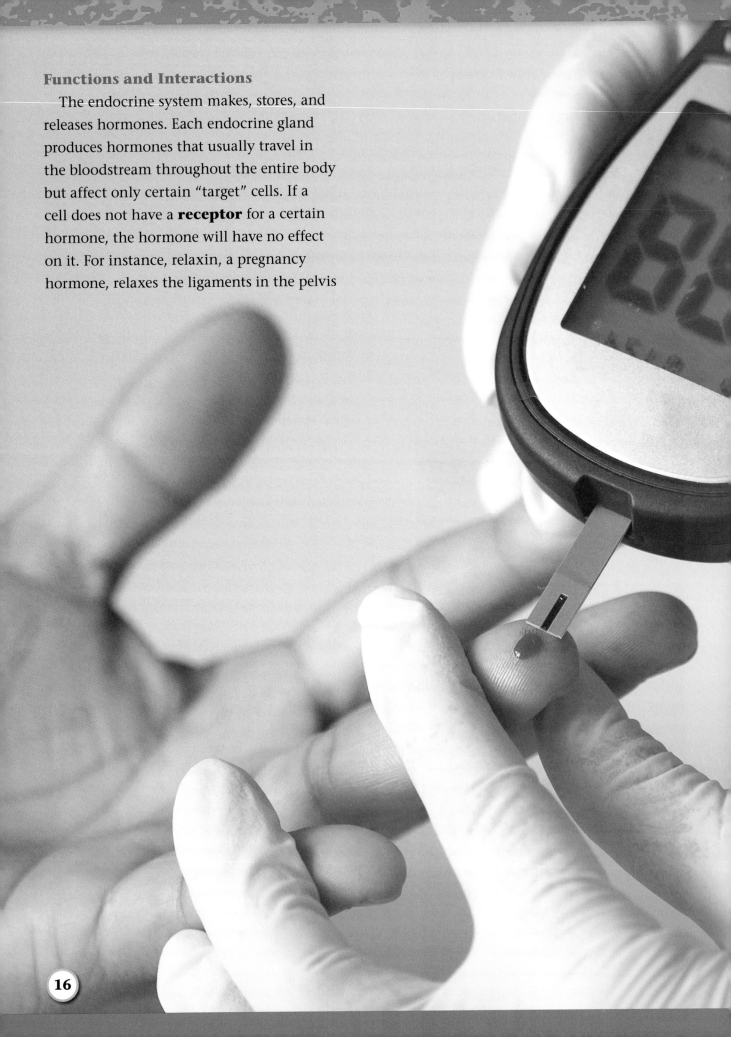

Functions and Interactions

The endocrine system makes, stores, and releases hormones. Each endocrine gland produces hormones that usually travel in the bloodstream throughout the entire body but affect only certain "target" cells. If a cell does not have a **receptor** for a certain hormone, the hormone will have no effect on it. For instance, relaxin, a pregnancy hormone, relaxes the ligaments in the pelvis

Growth and development. The pituitary gland has a big effect on many organs. It produces human growth hormone, which controls how fast and how much a child's bones grow. Hormones also control male and female characteristics. In men, for example, they control the development of facial hair.

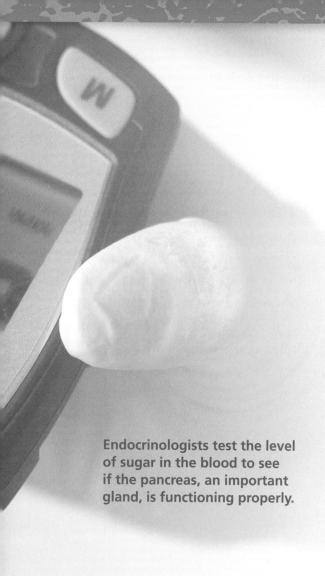

Endocrinologists test the level of sugar in the blood to see if the pancreas, an important gland, is functioning properly.

Young men develop facial hair at puberty because of the hormone testosterone.

in preparation for childbirth. Cells in other ligaments in the body also have receptors for relaxin, making **joints** less stable and more prone to injury. Relaxin does not affect muscle or bone cells, only the cells in ligaments.

The endocrine system affects all the other organ systems. The hypothalamus receives information from all parts of the body. It sends hormonal instructions to the pituitary gland, which influences many other endocrine glands. The blood is the pathway that hormones use to reach both the other glands and other cells in the body.

Reproduction. The pituitary gland secretes a hormone called follicle-stimulating hormone (FSH). In women, FSH regulates egg release and menstruation. In men, FSH controls sperm production. In both men and women, the release of FSH is affected by the levels of other hormones in the bloodstream.

Homeostasis. Hormones control **homeostasis**, the internal balance of body systems. They regulate blood pressure. They help regulate water and body temperature. With the excretory system, they control the amount of water released from sweat glands and from the bladder.

Bone and muscle strength. Hormones released by the pituitary gland help adults maintain healthy muscle and bone mass.

Metabolism. Metabolism is a collection of chemical processes that converts the food we eat into energy. It occurs in your cells and is controlled by hormones. A hormone produced in the thyroid gland helps determine the speed of chemical reactions. Your pancreas secretes an important hormone called insulin. Insulin controls your cells' ability to absorb sugars from the bloodstream.

Response to stimuli. Hormones regulate how you feel **pain** and react to stress. They also control your emotions.

The hormone adrenaline is involved in the fight or flight response. In response to danger, the brain sends a signal to the adrenal glands to release adrenaline. Adrenaline enters the bloodstream and is carried around the body, affecting target cells in the heart, lungs, and skin. It increases heart rate, blood pressure, and breathing rate, and causes the hair on your arms to rise. It directs blood flow to major **skeletal muscles**. During stressful moments, adrenaline even slows down the digestive system and heightens your senses. Your body becomes totally ready to respond to danger.

Many people crave the heart-pounding adrenaline rush that comes with an exciting challenge.

When Things Go Wrong

There are many types of endocrine disorders. These symptoms can mean that something is wrong with the endocrine system.

- Abdominal **pain**
- Dehydration or excessive thirst
- Nausea
- Hyperactivity
- Fatigue
- Depression
- **Abnormal** weight gain or loss

Excretory System

Your body is full of poisons. Every second, digestion and the activity inside your **cells** produce harmful wastes. These products can damage or even kill cells if they aren't removed.

The kidneys, urinary tract (urethra, kidneys, ureter, and bladder), lungs, liver, large intestine, and skin make up the excretory system. This system removes toxic wastes from the body. The excretory system is also responsible for maintaining the correct balance of water and salts in your cells.

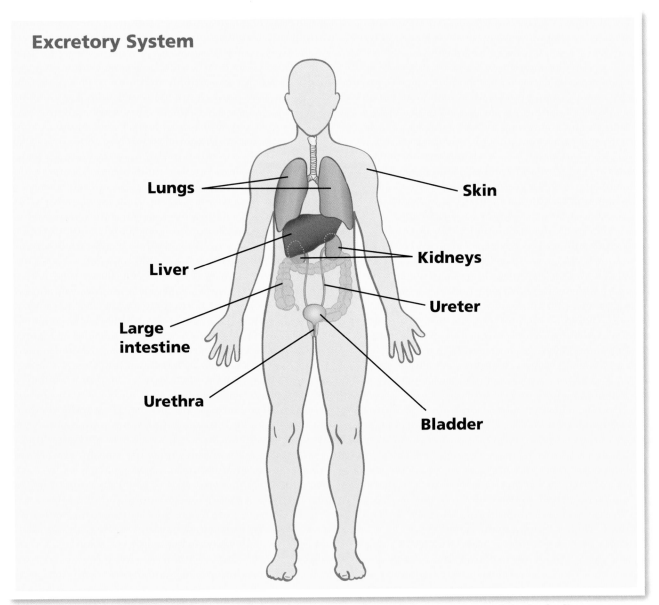

Excretory System

Lungs

Skin

Liver

Kidneys

Ureter

Large intestine

Urethra

Bladder

The excretory system is made up of organs that remove harmful waste materials, including organs that we usually consider part of other systems, such as the lungs and large intestine.

What you put into your body affects the way it comes out. Fruits like blueberries, along with fluids and fiber-rich grains and vegetables, help the large intestine do its job of removing solid waste.

Parts of the Excretory System

Kidneys and urinary tract. The kidneys are a pair of bean-shaped organs located in your abdomen along either side of the spine. They are the most important organs in the excretory system. Kidneys filter waste products from your blood. Every hour, the blood in your body passes through the kidneys about 12 times. In the process, the kidneys produce urine, one of the main products of the excretory system. Because they are so important, the kidneys are protected by the rib cage and are surrounded by a layer of fat.

The ureters are thin tubes of muscle. They carry urine to the bladder, a stretchy muscular organ. The bladder stores urine and releases it through a tube called the urethra. This process is called urination.

Lungs. The lungs are spongy air-filled organs located on either side of your chest. They are part of the respiratory and excretory systems. As part of the excretory system, they help remove carbon dioxide from your body.

Liver. The liver is the second largest organ in your body. It weighs about 1.5 kilograms (kg). You cannot live without a liver. It breaks down harmful substances before they get to other parts of your body.

Large intestine. The large intestine is about 3 m long. It is also part of the digestive system. As part of the excretory system, it reclaims usable water and removes leftover solid waste (feces). Did you know that the average person gets rid of about 140 kg of waste every year?

Skin. The skin is the largest organ of your body. The skin of an average adult weighs about 3.5 kg. Sweat glands in the skin release waste materials such as salt. These glands play only a minor role in the excretory system.

Sweat glands eliminate a salty fluid through pores in the skin. Sweating also cools the body.

Every cell, tissue, and organ in your body depends on water. In fact, water makes up around 60 percent of your body weight.

Functions and Interactions

Water and mineral balance. Your cells would swell and burst if they had too much water. With too little water, they could shrivel and die. The excretory system helps maintain the right balance of water and salts in your body.

This balance is controlled by hormones produced by the endocrine system. For instance, if you drink too much water, hormones signal the kidneys to produce more urine. If you haven't been drinking enough water, hormones signal the kidneys to absorb water into the blood. When this happens, your urine becomes darker than usual because it contains less water.

The excretory system helps regulate blood pressure. It also helps regulate the level of minerals such as calcium, potassium, and sodium. These minerals are important in regulating heartbeat.

Waste removal. The excretory system breaks down and eliminates cells' waste products. We often focus on what is coming into our bodies and take for granted what gets rid of waste. What would happen to our bodies if waste just kept building up inside?

For example, carbon dioxide is a waste produced in your cells. High levels of carbon dioxide can cause headaches, convulsions, unconsciousness, and even death. But there is a way to get rid of the gas. Blood carries carbon dioxide from the cells to the lungs. The lungs release carbon dioxide each time you exhale.

Another example is water. Our bodies need water. But water is also a waste product. Excess water travels from cells to the kidneys and then to the bladder. As the bladder fills up, stretch **receptors** (nerve cells) send signals to the brain. The brain responds by sending a signal to urinate. The nervous system then works with the muscular system to control the muscles in the urethra. Babies have no conscious control over these muscles. When the bladder is full, they empty it (thank goodness for diapers!). With training, we learn to control these muscles, even in our sleep.

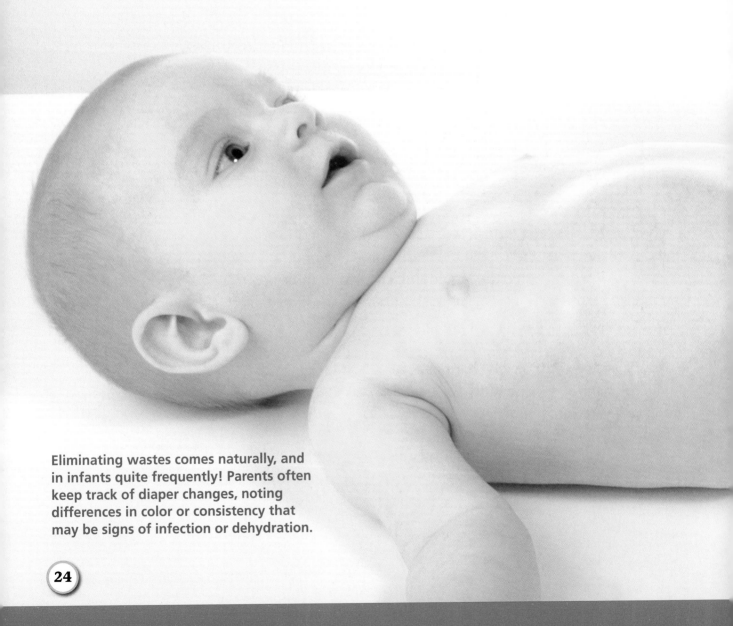

Eliminating wastes comes naturally, and in infants quite frequently! Parents often keep track of diaper changes, noting differences in color or consistency that may be signs of infection or dehydration.

Another dangerous waste is ammonia. When the digestive system breaks down proteins, it produces ammonia. The liver converts ammonia into urea. Urea dissolves in urine and safely passes from the body.

When Things Go Wrong

These symptoms can mean that something is wrong with the excretory system.

- Abdominal **pain**
- Intense pain in the side and back, below the ribs
- Pain while urinating
- Color in the urine
- Inability to urinate
- Nausea and vomiting
- Fever and chills

Muscular System

When you look at other people's faces, can you tell what they are feeling? Facial expressions are the result of the movement of 36 different facial muscles. A muscle that runs from your forehead back over your scalp, the epicranius, allows you to wrinkle your forehead. Another is called the "frowning" muscle and causes the vertical wrinkles above the nose. Sometimes we control those muscles (give your partner a fake smile). And sometimes our facial expressions are involuntary. What facial expressions do you make if you are concentrating, surprised, or angry? Those usually involve involuntary facial expressions.

The muscular system allows your body to move. Muscles move the skeleton so we can run, write, talk, and so on. Muscles also move things along inside your body.

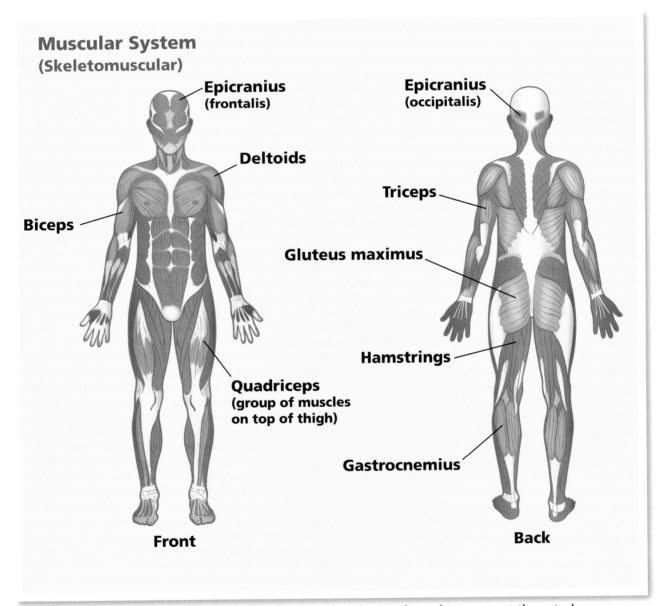

Muscular System (Skeletomuscular)

Epicranius (frontalis)

Epicranius (occipitalis)

Deltoids

Triceps

Biceps

Gluteus maximus

Quadriceps (group of muscles on top of thigh)

Hamstrings

Gastrocnemius

Front

Back

The muscular system consists of skeletal muscles and the tendons that connect them to bones. Every move you make is the result of muscle action.

Muscles often work in pairs: one contracts or flexes while the other extends or relaxes. Key muscle pairs in weightlifting are the quadriceps and hamstrings in the thighs.

Parts of the Muscular System

You have nearly 700 muscles in your body. In total, they make up almost half your weight. The muscular system has three types of muscle tissue: skeletal, smooth, and cardiac. Each type has slightly different versions of muscle **cells**. Muscle cells link up to form the specific type of muscle tissue.

Skeletal muscle. When you say *muscle*, you usually mean **skeletal muscle**. Every voluntary physical action uses skeletal muscle. Each skeletal muscle is a single organ made of muscle tissue, blood vessels, **tendons** (which connect muscles to bones), and nerves. The cells of skeletal muscles link into strong, straight muscle fibers. The biceps and triceps in the arm, the quadriceps and gastrocnemius in the leg, and the deltoids in the shoulder are all examples of skeletal muscle.

Smooth muscle. Smooth muscle is found inside organs such as the stomach, intestines, and blood vessels. Smooth muscle looks smoother under a microscope than the other two types of muscle. Smooth muscles are called involuntary muscles. They are automatically controlled by your brain, which means that you can't tell your stomach to stop digesting food (or growling!).

Cardiac muscle. The heart is made up of **cardiac muscle**. It is involuntary muscle that pumps blood through the body. Under a microscope, it has dark and light stripes. The cells of cardiac muscle are X- or Y-shaped and interconnected. These characteristics make the tissue strong enough to last a lifetime.

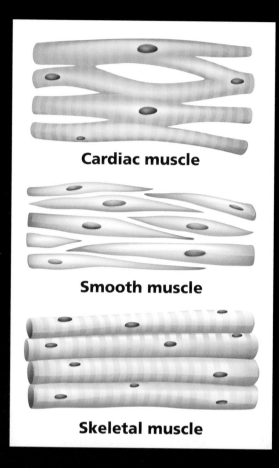

Cardiac muscle

Smooth muscle

Skeletal muscle

Because of their rigorous training methods, gymnasts have very defined skeletal muscles.

Functions and Interactions

The muscular system works closely with the nervous system in both voluntary and involuntary ways. You don't have to think about the **smooth muscles** while breathing or moving food through your digestive tract. But raise your hand or make a fist. This movement is controlled by your voluntary skeletal muscles.

Every cell in every kind of muscle tissue needs oxygen and nutrients to function. The respiratory system provides oxygen to the muscle cells. The digestive system breaks down food into nutrient molecules (like sugars) that all cells need. The circulatory system carries nutrients and oxygen to the muscle tissue. It carries waste products like carbon dioxide and lactic acid from the muscle tissue to the excretory system.

Body movement. Straighten your leg out. As you do, the skeletal muscles at the front of your thigh contract while the muscles at the back of your thigh relax. Your brain sends a signal to move the muscles.

A single skeletal muscle often attaches to two different bones. When the muscle flexes and shortens, it pulls the bones closer together. These contractions move **joints** like the elbow or knee. Skeletal muscles usually work in groups to produce precise movements. You probably use around 200 muscles just to take one step.

Muscles are even connected to the eyeball! The eyelid muscle is the fastest-reacting muscle in the entire body. It can contract in less than 0.01 second.

Moving materials through your body. As smooth muscle contracts, it moves things through your body. Food moves through the digestive system, waste through the excretory system, and blood through the circulatory system.

The digestive system has smooth muscle in the esophagus, stomach, and small and large intestines. Rhythmic contraction and relaxation of the muscles moves food through the entire digestive system. Smooth muscle also makes up the bladder and tubes of the excretory system. It allows you to control your urination.

Stretching through the full range of motion helps prepare muscles so they are ready for exertion. Lunges like these loosen the major muscle groups used in running.

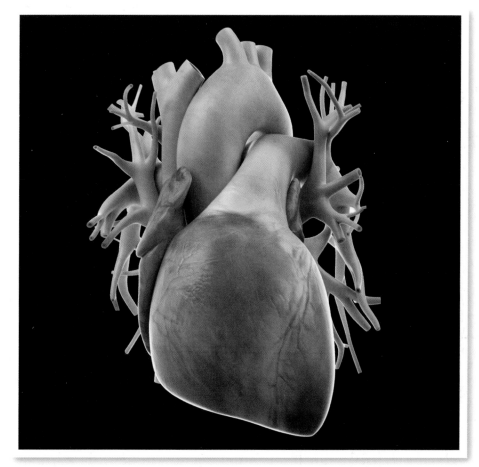

The powerful and continuous beating of your heart is an example of involuntary muscle action. That means you do not control, or even have to think about, the movement.

The cardiac muscle of the heart and the smooth muscle of the arteries regulate blood flow. Cardiac muscle causes the rhythmic beating of your heart. As the muscle contracts, blood pumps from the heart through the blood vessels to all the cells in the body. **Hormones** from the endocrine system and the brain can adjust the rate of cardiac muscle contraction. But electrical signals from the cardiac muscle cause the contractions. All the cells of cardiac tissue beat together as one.

At birth, you have all the muscle fibers you will ever have. The average heart will beat more than 2.5 billion times in a lifetime. It probably does the most work of any muscle in the body.

When Things Go Wrong

These symptoms can mean that something is wrong with the muscular system.
- Overall weakness
- Muscle spasms, cramping, and twitching
- Muscle aches, **pains**, and stiffness
- Muscle weakness or slackness
- Paralysis
- Difficulty swallowing
- Difficulty breathing
- Drooping eyelids
- Double vision

Nervous System

You are in the school cafeteria. Suddenly out of the corner of your eye, you see a piece of food flying toward your head! You duck and look around, trying to figure out who threw the food. The nervous system is involved in all those actions. Your eye senses movement. That information travels to your brain. Your brain causes your muscles to respond by ducking to avoid the food. Your brain also causes you to look for the source of danger.

The brain, **spinal cord**, sense organs, and **neurons** make up the nervous system. The nervous system monitors and responds to the outside environment. It also controls all the activity inside the body.

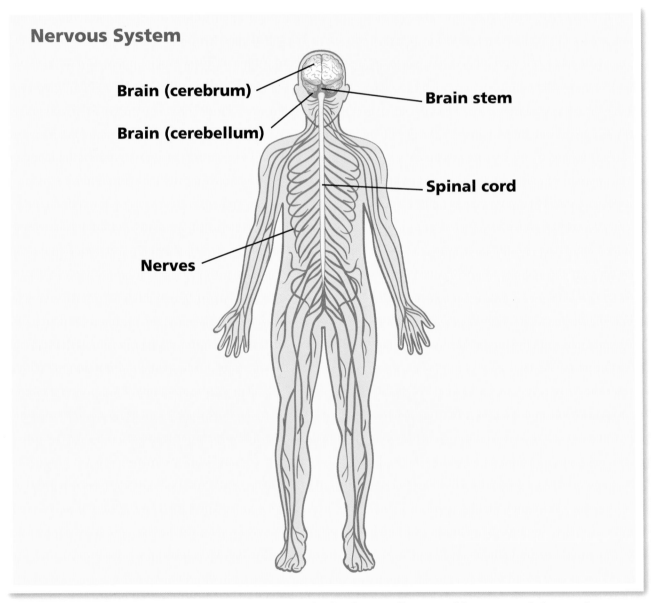

Nervous System

Brain (cerebrum)

Brain (cerebellum)

Brain stem

Spinal cord

Nerves

The nervous system handles communication in the body. It collects and interprets information from other systems and from outside surroundings and directs the body's responses.

All the learning a student does is possible because of the brain—reasoning, remembering, solving problems, sharing ideas, even imagining. So, in a way, every student is brainy.

Parts of the Nervous System

The nervous system has two parts. The **central nervous system** is made up of your brain and spinal cord. The **peripheral nervous system** consists of all the other nerves. The peripheral nervous system receives **stimuli** from **sensory receptors** and sends them to the brain. The receptors constantly monitor the body's internal and external conditions.

Brain. The brain is a soft, wrinkly organ, protected by the bones of the skull. It weighs about 1.5 kg. It is made up of 100 billion nerve **cells** called neurons. The brain allows you to experience everything around you, storing some of those experiences as memories in the cerebrum. The cerebrum controls your thinking, planning, and voluntary actions (like deciding to talk to a friend). The brain stem controls involuntary actions (like breathing, heartbeat, and digesting food). The cerebellum is involved in balance, coordination, and learning complex actions, like how to hit a baseball.

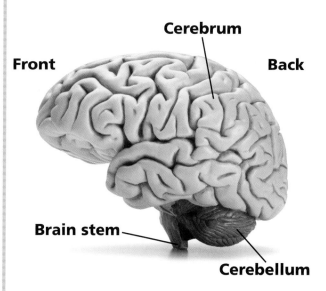

Cerebrum

Front   **Back**

Brain stem

Cerebellum

The brain, sometimes called gray matter, is the control center of the nervous system.

Spinal cord. The spinal cord is a long, thin bundle of neurons. It carries information back and forth between the brain and the other parts of the body.

Sense organs. The sense organs are the eyes, nose, ears, skin, and tongue. These organs have sensory receptors that connect to neurons. The neurons send the information to the brain to be processed and, sometimes, acted upon.

Neurons. Neurons exist in almost every part of your body. They transmit signals using electricity and chemicals. **Sensory neurons** send information from sensory receptors in the sense organs to the central nervous system. **Motor neurons** send information the other way, from the central nervous system to muscles and glands. **Interneurons** connect the sensory and motor neurons.

Up to 5 trillion nerve impulses per second travel along cells called neurons through the body to the brain, where information is processed.

Functions and Interactions

The nervous system is like a communications highway, carrying messages to and from the brain. The nervous system enables you to think, learn, breathe, and respond to the environment. You can read these pages because many elements of your nervous system are working together.

Sensory function. As you interact with the world, your sense organs relay information to specific areas of your brain. Vision, balance, hearing, touch, smell, and taste receptors collect information. Your brain interprets all that information.

Motor function. Pick up a pencil. This simple task requires complex decisions and muscle movement. Your brain figures out where the pencil is. It decides which muscles to contract to pick up the pencil. And it determines how much force is necessary to pick it up. Every muscle depends on signals from the nervous system.

Processing and interpreting information. Once information has traveled to your brain, the brain processes it for immediate action, learning, or memory storage. The **autonomic nervous system** is part of the peripheral nervous system. It controls the functions of internal organs. Some functions, such as your breathing, heartbeat, and digestion, are controlled unconsciously. You don't need to think about digesting after lunch or breathing when you are asleep.

All organs require instruction and direction from the nervous system. So the nervous system is constantly interacting with every other organ system. The nervous system controls the flow of blood through the circulatory system. At the same time, the brain depends upon constant blood flow for oxygen from the lungs and nutrient molecules from the digestive system.

Because the nervous system is so important, other parts of the body work to protect it. For example, two parts of the skeletal system (the skull and vertebrae) protect the main organs of the nervous system (the brain and spinal cord). Many major nerve bundles are protected by **joints** or hidden beneath bones or major muscles. The skeletal system also stores calcium, which is important for nerve functions.

The simple task of picking up a pencil is not so simple after all. It requires coordinated decisions and sensations, precise muscle activity, eye-hand coordination, and more.

You control what you eat and drink, but then your nervous systems takes over, directing the digestive and excretory systems to extract nutrients and eliminate wastes.

The endocrine system works closely with your brain and spinal cord to control the production and release of hormones.

Your nervous system unconsciously controls most of the actions of your digestive and excretory systems. You consciously control some of those actions, however, such as when to eat, drink, and urinate.

There is much we know about how the brain works and much still to find out. Mysteries remain about how memories are formed and stored, differences between the brains of men and women, and how learning occurs.

When Things Go Wrong

These symptoms can mean that something is wrong with the nervous system.

- Headaches
- Dizziness, **fatigue**, and fainting
- Memory loss and confusion
- Nausea (with or without vomiting)
- Weakness and numbness of parts of the body
- Behavioral changes
- Incoherent speech
- Muscle twitching, spasms, and seizures
- Extreme sleepiness and inability to wake up
- Inability to sleep

Respiratory System

Are you breathing? Did you have to think about it? You can control your breathing. But most often, breathing in and out is unconscious and controlled by your nervous system. How many breaths do you take in a minute? For a human, the average range is between 12 and 20. For a giant tortoise, the rate per minute is about 4, while for a mouse it can be as high as 160 breaths per minute.

Without oxygen, **cells** cannot survive, so you cannot survive. The respiratory system is the organ system that takes in vital oxygen and removes potentially toxic carbon dioxide.

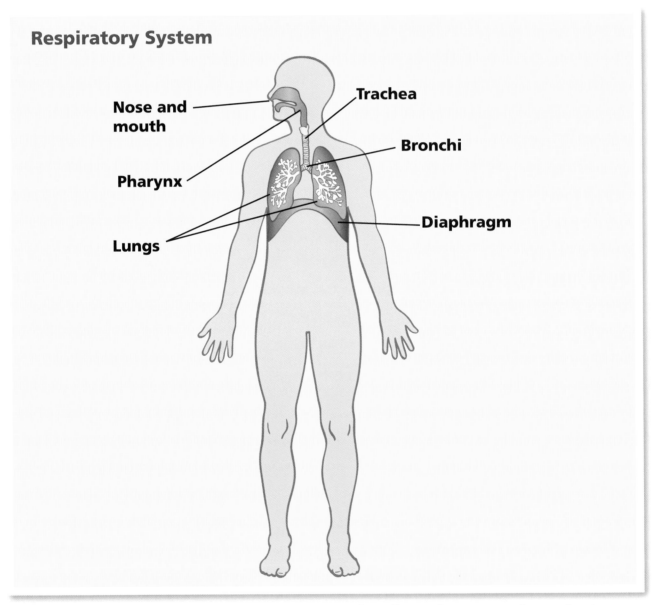

Respiratory System

Nose and mouth

Pharynx

Lungs

Trachea

Bronchi

Diaphragm

The respiratory system handles the body's oxygen supply. Cells need oxygen to change food into energy. They use up oxygen and give off carbon dioxide in a process called aerobic cellular respiration.

Parts of the Respiratory System

The respiratory system has three main parts: the airway, the lungs, and the muscles that cause the lungs to inhale and exhale air.

The airway. The airway carries air to and from the lungs. The airway includes the nose, mouth, pharynx (the top of the throat), trachea (windpipe), and bronchi. The nose and mouth bring air in. The nose filters and warms air as it enters the pharynx. The air moves past your pharynx and into your trachea. The trachea is made of stiff **cartilage**, which keeps the airway open.

It branches into two bronchi, which carry the air to the left and right lungs.

Lungs. The lungs are spongy organs of connective tissue with 2,400 kilometers (km) of airway passages. That's the distance from San Francisco, California, to Kansas City, Missouri! Each lung contains on average 300 to 500 million tiny air sacs called **alveoli**. This is where blood exchanges gases—oxygen and carbon dioxide. The lungs are very important organs. They are protected by the rib cage.

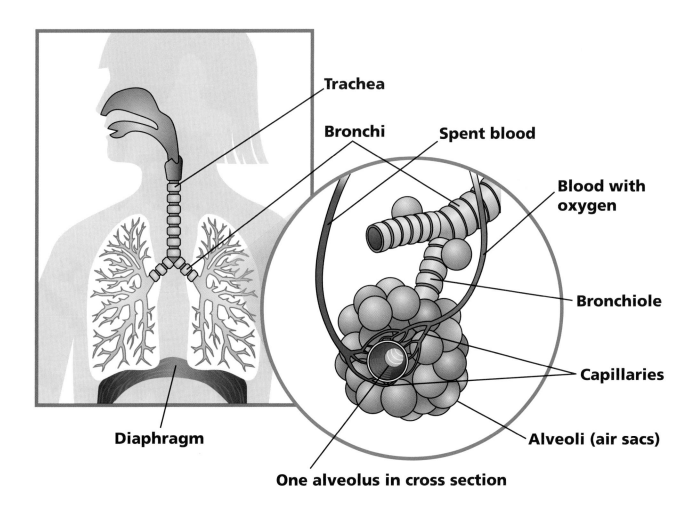

Trachea

Bronchi

Spent blood

Blood with oxygen

Bronchiole

Capillaries

Alveoli (air sacs)

Diaphragm

One alveolus in cross section

The airway passages end as millions of tiny alveoli covered by capillaries.

Respiratory muscles. Muscles help move air all along the respiratory tract. **Neurons** in the brain stem automatically send signals to the diaphragm and rib muscles. They contract and relax at regular intervals. This muscle movement causes the lungs to expand and contract. What happens when you inhale? Your rib muscles expand, and the diaphragm muscle (located underneath the lungs) contracts. The lungs expand and fill with air. When you exhale, the diaphragm muscle relaxes. Your rib muscles contract to push air out of the lungs.

Functions and Interactions

Each cell in your body needs a constant supply of oxygen to produce usable energy. The cells also need to get rid of carbon dioxide, a toxic waste. Exchanging these gases is the work of the respiratory system.

When you exhale, you expel not only carbon dioxide waste but also water vapor. On a cold day, the warm water vapor condenses into tiny water droplets, and you can "see your breath."

Providing oxygen. When air enters the lungs, it flows through increasingly smaller airways to the alveoli. There, oxygen transfers to capillaries, the tiny blood vessels of the circulatory system. Oxygen then diffuses into the bloodstream. The oxygen bonds to hemoglobin molecules in the red blood cells. The heart pumps the oxygenated blood to the cells in the body.

Sometimes you have trouble breathing before air even gets to the lungs. The **epiglottis** is a flap that directs food down a tube called the esophagus and air down the trachea. If the epiglottis does not close all the way when you swallow food, some may go down the "wrong pipe" and cut off the air supply. Choking and coughing are involuntary muscular responses that try to clear any food blocking the trachea.

When you exhale sharply, such as to blow on a dandelion, your diaphragm relaxes and your rib muscles contract quickly.

Swimmers need to exhale and inhale to maximize air volume and the speed of gas exchange.

Removing waste carbon dioxide.
The lungs can also be considered part of the body's excretory system. Blood containing waste carbon dioxide from your cells travels to the alveoli. There, the carbon dioxide moves out of the **capillaries** and into the alveoli air space. It leaves your body when you exhale.

Without thinking, you breathe more than 19,000 times per day. When you hold your breath for any length of time, you start to feel burning in your lungs and have to take a breath. It is not the need for oxygen that triggers the reflex, but the increasing levels of carbon dioxide, which can be deadly. The brain senses the increasing levels and signals the diaphragm and muscles between your ribs to spasm, forcing you to gasp, exhaling the extra carbon dioxide and bringing in fresh oxygen.

Another cellular waste product is not so toxic. Put your hand in front of your mouth and exhale. What do you feel? Any moisture you feel is water vapor.

Asthma is a condition in which the smallest airway passages swell and narrow, making breathing difficult. Inhalers work to get medicine directly into the lungs to help stop these breathing problems.

When Things Go Wrong

These symptoms can mean that something is wrong with the respiratory system.

- Chest **pain**
- **Fatigue**
- Difficulty breathing
- Persistent coughing or sneezing
- Rapid, shallow, or **abnormally** deep breathing
- Periodic absence of breathing (known as sleep apnea)
- Fever

Skeletal System

Who has more bones—You? Your teacher? A newborn baby? Or do you all have the same number? Take a guess, and we'll get back to the answer in a moment.

Bones, **cartilage**, and **joints** make up the skeletal system. This system has several major functions, including support, movement, protection, production of blood **cells**, and storage and regulation of **hormones.**

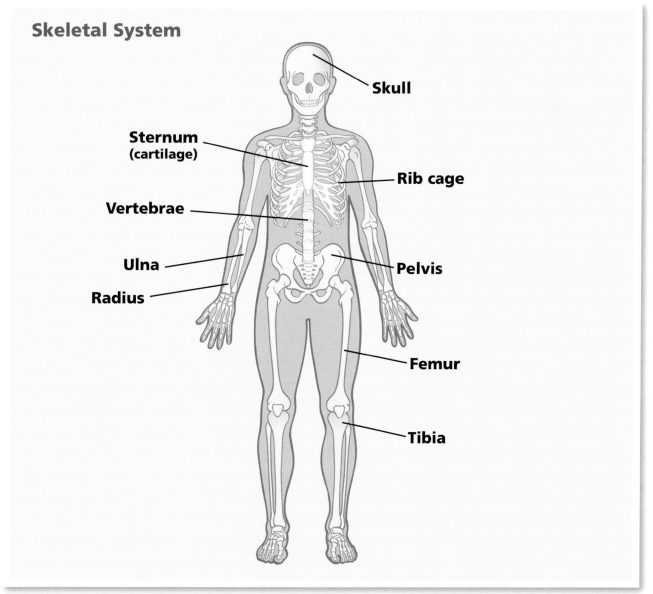

Skeletal System

Skull

Sternum (cartilage)

Rib cage

Vertebrae

Ulna

Pelvis

Radius

Femur

Tibia

The skeletal system consists of the bones and cartilage that support, protect, and give structure to the body. Joints, like elbows and knees, are places where bones meet.

Parts of the Skeletal System

Bones. Every bone in your body is a separate organ made up of cells and tissues. Your bones are alive. Bone building goes on throughout your life. Special bone cells called **osteoblasts** are responsible for making new bone tissue and repairing damage.

Passing in and out of all bones are many blood vessels. They bring oxygen, nutrient molecules, and minerals to bone tissue. Blood vessels carry away new blood cells that are made in **bone marrow**.

Joints. Joints are where bones meet. The type of joint determines the possible kinds of movement. For example, the joints of your wrist allow the bones in your wrist to move in several ways. The joint that connects the humerus in the upper arm to the radius and ulna in the forearm moves back and forth like a hinge. Joints are made up of cartilage and other types of connective tissue.

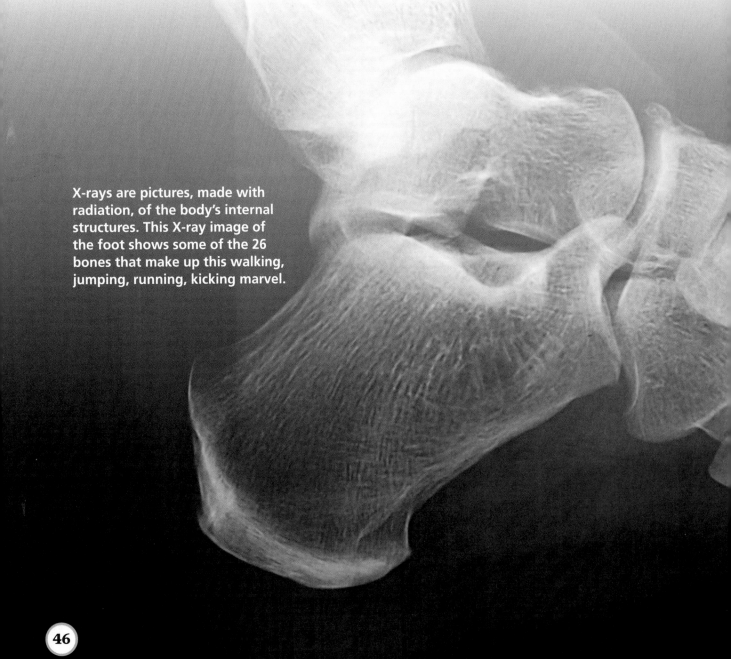

X-rays are pictures, made with radiation, of the body's internal structures. This X-ray image of the foot shows some of the 26 bones that make up this walking, jumping, running, kicking marvel.

Cartilage. Cartilage is a connective tissue. It is found in joints between bones and makes up parts of your ears, your nose, and the sternum in your rib cage. Its main function is to cushion the bones and reduce friction in joints. Cartilage isn't as hard as bone, but it isn't as flexible as muscle. Unlike bone, it doesn't contain blood vessels. As a result, it takes a long time to heal after an injury.

Teeth. Teeth are part of your skeletal system, but they aren't made out of bone and aren't counted as bones. They are made of several tissues, including enamel. Tooth enamel is thought to be the strongest substance in your body. Teeth are also considered part of the digestive system.

Let's get back to the earlier question: Who has more bones? A newborn baby! At birth you have about 270 bones. Over time, some of those bones fuse. By adulthood, you have about 206 bones.

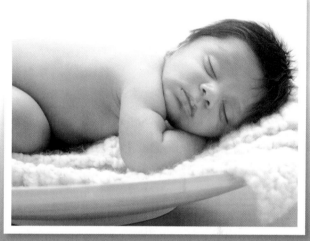

A newborn has more bones than you do. Some fuse, or grow together, as the baby develops.

Functions and Interactions

Support. Stop and think what we would look like if we didn't have a skeleton. Most likely, we would be blobs. The skeleton gives us shape. Your skeleton determines your height and how big your feet and hands will grow to be. The skeleton helps support the body and helps keep internal organs, such as the brain, in place.

Movement. Bones such as the femur and tibia in the leg are the attachment points for muscles. They move when the muscles attached to them contract. Without solid bones and joints, muscles would not be able to move the body. The nervous system initiates and coordinates this movement.

Protection. Bones protect many of the organs in other organ systems. The skull protects the brain. The bones in your rib cage protect the lungs and heart. The vertebrae in your back protect the **spinal cord**. Pelvic bones protect the lower part of your intestines and your urinary tract, which is part of the excretory system.

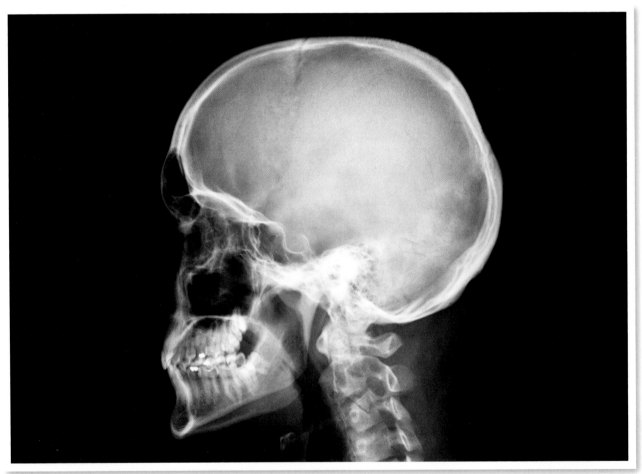

The skull is actually a collection of bones. Some form the cranium, a thick, round helmet that protects the brain. Others, like the movable jaw bone, support the muscles and organs of the face.

Bone Structures and Tissues

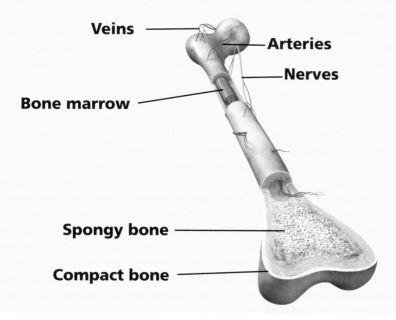

- Veins
- Arteries
- Nerves
- Bone marrow
- Spongy bone
- Compact bone

Calcium makes bones hard and strong, but only on the outside. Inside many bones, including the femur, is soft tissue called bone marrow, which makes blood cells and stores fat.

Production of blood cells. The center of most bones is filled with a soft spongy material called bone marrow. The marrow contains cells that produce all the body's **platelets**, **red blood cells**, and some **white blood cells**. When you lose blood or your liver recycles old blood cells, the replacements come from inside your bones.

Storage. Bones store minerals. For example, bones store calcium and release it into the bloodstream when needed by your body. Your bones can also store iron.

Regulation of hormones. Insulin is a hormone in the endocrine system that regulates blood sugar levels and fat storage. Bone cells release a protein that regulates insulin production and effectiveness.

When Things Go Wrong

These symptoms can mean that something is wrong with the skeletal system.

- Joint **pain**
- Back pain
- Soreness in the hands or feet
- Swelling near a joint

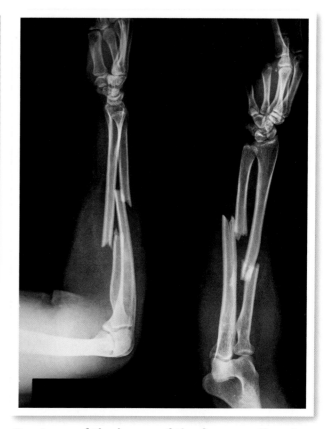

Fractures of the bones of the forearm, the ulna and radius, are the most common bone breaks.

Aerobic Cellular Respiration

No living thing can survive without energy. The trillions of cells in your body need energy to grow, reproduce, and respond to the environment.

All multicellular organisms, including plants, mushrooms, insects, and snails, must supply every cell with energy. And every single-celled organism, such as bacteria, must obtain energy.

Humans get energy from food. Food molecules, such as **glucose**, store potential energy in their chemical bonds. Cells can't use that potential energy directly. It is gradually released during a series of chemical reactions that require oxygen. Those reactions are **aerobic cellular respiration**. Here is the chemical summary.

$$6O_2 + C_6H_{12}O_6 \rightarrow 6CO_2 + 6H_2O + \textit{usable energy}$$

oxygen glucose carbon water
dioxide

Food tastes good, but that's not really why we eat. We get energy to live from the food we eat.

Fats and proteins are also sources of energy-rich chemical bonds. Fats are used primarily for energy. Proteins provide raw materials for making muscle tissue, hormones, and other compounds and tissues. They can also be used for energy if sugar or fat is not available.

Aerobic cellular respiration occurs in every cell of almost every living thing. How do cells get the oxygen and glucose needed for aerobic cellular respiration?

Breathing and aerobic cellular respiration are related, but they are not the same.

Insects, mushrooms, and every other living thing need energy to carry out life processes. The energy is released and used in the organism's cells.

Oxygen

Humans breathe in oxygen from the atmosphere. Blood picks up the oxygen in our lungs and carries it to the cells. There it becomes available for aerobic cellular respiration.

Plants don't breathe. So how do they get oxygen? Plants produce oxygen as a waste product during **photosynthesis**. Plant cells use some of that oxygen to perform aerobic cellular respiration.

Did You Know?

Some single-celled organisms live in places without oxygen. They use fermentation, which does not need oxygen, to produce energy. Our cells can use this process, too. It kicks in during intense exercise. But fermentation does not provide as much usable energy as aerobic cellular respiration.

Why is running an aerobic exercise? Because more physical activity requires more energy, and more energy requires more oxygen getting to the cells.

Photosynthesis

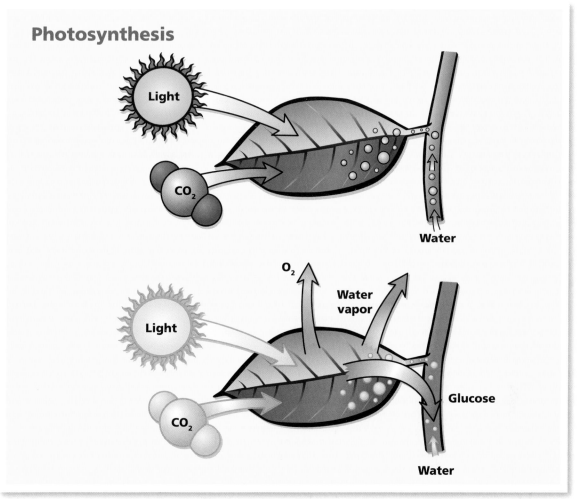

Plants use light energy from the Sun to turn carbon dioxide and water into sugar. Plants use the sugar for food and can use some of the oxygen for aerobic cellular respiration. The oxygen not used by the plant cells is released into the air.

Sugars

Some organisms can make their own glucose. They capture energy from the Sun, using a substance called chlorophyll. They use the Sun's energy to turn carbon dioxide gas and water into energy-rich sugars and oxygen. This process is photosynthesis. Here is a chemical summary of photosynthesis.

$$6CO_2 + 6H_2O + \textit{light energy} \rightarrow C_6H_{12}O_6 + 6O_2$$

carbon dioxide   water   glucose   oxygen

The energy that originated as solar energy is now stored in the bonds of glucose molecules. It can be released as usable energy by aerobic cellular respiration. The cells of the photosynthetic organism can use this chemical energy to run their own activities.

Photosynthesis occurs mainly in the green leaves of plants.

Humans can't make their own food. We eat plants or other animals that have eaten plants. Our digestive system breaks down food into sugar, protein, and fat molecules. Those energy-rich molecules move into the bloodstream from the small intestine, and travel to the cells. There, the food molecules become available for aerobic cellular respiration.

The cells break the bonds in the sugar molecule. Energy transfers to a usable form that cells (and consequently the body) can use to perform all the work they need to do.

Products

What are the products of aerobic cellular respiration? Carbon dioxide and water. These products leave the cells and are transported out of the body.

Summary

Aerobic cellular respiration is critical for life on Earth. The energy obtained from the breakdown of sugars, such as glucose, that are made by photosynthesis is what runs our cells and our bodies.

Think Question

A student said, "Plants do photosynthesis. Animals do aerobic cellular respiration." Do you agree or disagree with this statement? Explain your thinking.

Comparing Processes

	Aerobic cellular respiration	Photosynthesis
Occurrences	In cells of all organisms	Only in plants, algae, and some bacteria
Function	To break down food, making energy available to cells	To transfer energy from the Sun to chemical bonds
Reactants	Glucose and oxygen	Carbon dioxide, water, and light energy
Products	Carbon dioxide, water, and usable energy for cells	Glucose and oxygen

Sensory Receptors

Mmm, I smell doughnuts! I can picture which one I want before I even see them. Why does that smell bring up images and memories in my brain?

One of the jobs of your nervous system is to take in information, decide what it means, and respond to it in some way.

For many, the smell of doughnuts means something tasty. The brain compares the smell to prior experiences. The smell stimulates a memory of a doughnut, a warm, chocolate treat. The brain recognizes desire and responds by sending the body to find the doughnut and eat it.

Our senses all rely on the ability of **sensory receptors** to take in information and translate it into electric impulses. These impulses travel to the brain. Each sense has receptors that respond to certain inputs: mechanical, chemical, or **electromagnetic**.

We experience the world around us through our senses, which work together to collect information and send it to the brain. How might all five senses help you enjoy a doughnut?

Mechanoreceptors

You investigated mechanical inputs in the **sense of touch**. Your skin has several types of pressure receptors that respond to touch. Each **mechanoreceptor** sent a message to your brain to tell it about the touch.

Other mechanoreceptors help you get to the bathroom in time. When your bladder is about half full, stretch receptors send signals along your pelvic nerves to your spinal cord.

A signal returns to muscles in your bladder, telling certain muscles to contract. Those contractions put pressure on the bladder. That pressure makes you want to urinate.

Think Question

Where else might mechanoreceptors be inside the body?

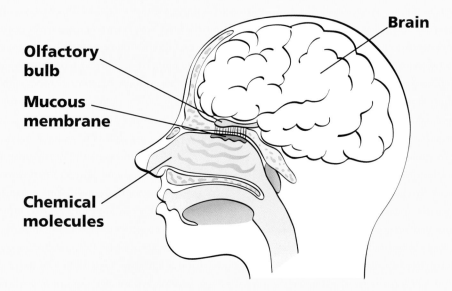

Sense of Smell

Brain

Olfactory bulb

Mucous membrane

Chemical molecules

Chemoreceptors in your nose work with your brain to figure out whether you are smelling flowers, food, or smoke from a fire.

Chemoreceptors

You learned about chemical inputs in your nose when you investigated the **sense of smell**. Chemicals in the air enter the nasal cavity as you inhale. The top of the nasal cavity is covered with mucus that dissolves the chemicals.

The human nose has about 400 types of **chemoreceptors**. Some scientists say these receptors can detect at least 1 trillion odors. Each chemical molecule fits into a special shape on a chemoreceptor. The chemoreceptor then sends an electric message to a **sensory neuron**. The **neuron** sends a message to the brain. The brain compares the message to previous experiences to determine what you smelled.

The **sense of taste** also uses chemoreceptors. Look at a partner (or in a mirror). Stick out your tongue. Notice all the little bumps on your tongue? There are taste buds along the edges of the bumps—about 10,000 taste buds total.

Some chemicals in food dissolve in saliva. They touch the tongue's taste buds. The sides of the taste buds have different chemoreceptors that respond to one basic taste. Those tastes are salty, sweet, sour, bitter, and umami. Umami (meaty) was the most recently discovered taste. Scientists think that there might be a sixth taste—for fats.

The taste-bud neurons send messages to the brain. The brain compares them to previous experiences to determine what was tasted. Your brain combines a food's taste, smell, and texture into the sensation of flavor. You can see how smell affects taste by plugging your nose the next time you eat something.

Think Question

Why can't you taste food very well when you have a cold?

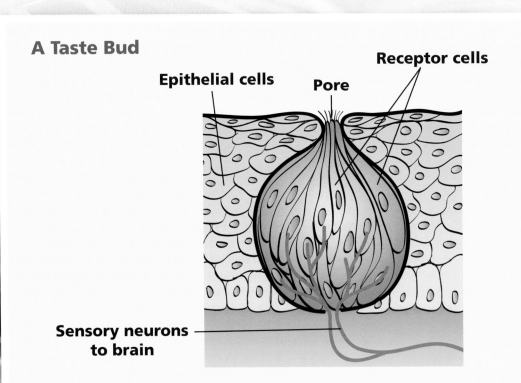

A Taste Bud

Epithelial cells

Pore

Receptor cells

Sensory neurons to brain

Taste buds have tiny receptor cells that respond to one of the five (maybe six) basic tastes. All tastes are combinations of these. It's up to the brain to recognize what you are eating.

Photoreceptors

You found that light is the **stimulus** for the **sense of sight**, or vision. Light is part of the electromagnetic spectrum. It comes in different wavelengths, which your visual sensory system perceives as different colors. Once the cornea and lens focus the light, the light forms an image on the retina. The retina contains **photoreceptors**, called **cones** and **rods**. Cones respond to color and detail. They are concentrated in the center of the retina. Rods respond to dim light and movement. They are most dense around the edges of the retina.

Each photoreceptor sends an electric message to the optic nerve. The optic nerve sends the message to the brain. The brain compares it to previous experiences to determine what you see.

The world of senses is amazing. Your brain processes all the information that comes in through all your senses, and your life is enriched. Imagine life without your senses!

Think Questions

1. **Which sense do you feel is most important? Why?**

2. **How does each of our senses function as a system? What are the inputs (stimulus) and the outputs (response)?**

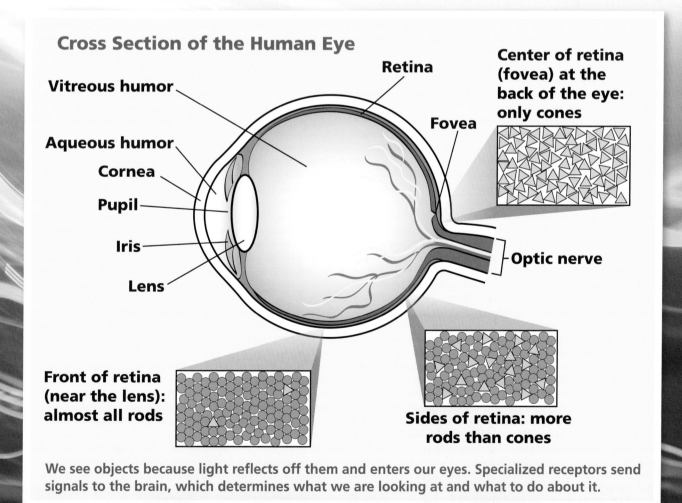

Cross Section of the Human Eye

Vitreous humor

Aqueous humor

Cornea

Pupil

Iris

Lens

Retina

Center of retina (fovea) at the back of the eye: only cones

Fovea

Optic nerve

Front of retina (near the lens): almost all rods

Sides of retina: more rods than cones

We see objects because light reflects off them and enters our eyes. Specialized receptors send signals to the brain, which determines what we are looking at and what to do about it.

Touch

Your sense of touch develops before you are born. It is the first sense to develop. Next to vision, touch is perhaps our most important sense for gathering information about the world.

Sense of touch includes several major types of sensation. Each type has one or more specific receptors. The receptor stimulus can be a mechanical or thermal signal.

Receptors

Pressure is a mechanical stimulation of the skin. Some pressure receptors respond to fluttering and light touch. Some respond to deep, steady indentation of the skin. Others respond to vibrations. Hair follicle receptors at the base of a hair signal slight movements of the hair.

Your fingertips are especially sensitive to touch.

Thermal receptors respond to sensations related to temperature. Cold receptors are no longer stimulated when the temperature drops too low, so hands full of ice may start to feel numb.

Pain and temperature signals are sent by bare nerve endings embedded in the skin. Some receptors sense hot or cold. Other bare-ending receptors sense sharp pain, deep aches, or damage caused by burning or freezing of skin.

Look at the diagram. Each receptor sends a message through nerves to the spinal cord. From there, the message travels to the brain. The brain processes the touch received on that part of the skin.

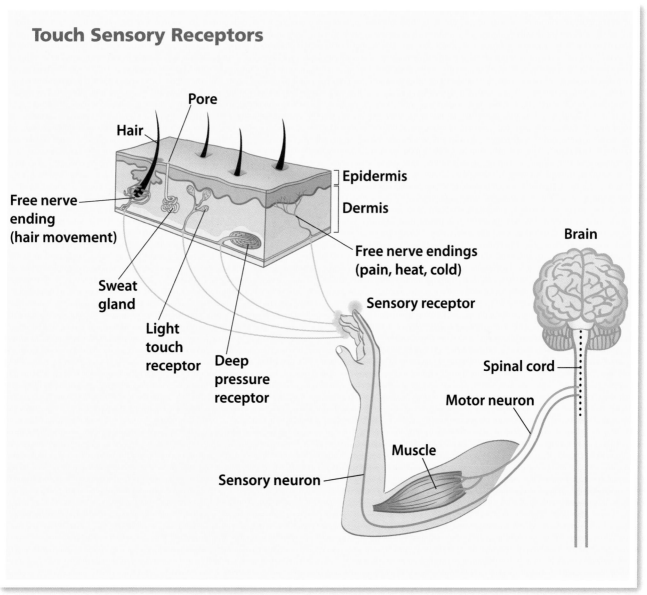

Touch Sensory Receptors

Pore

Hair

Epidermis

Dermis

Free nerve ending (hair movement)

Free nerve endings (pain, heat, cold)

Sweat gland

Sensory receptor

Brain

Light touch receptor

Deep pressure receptor

Spinal cord

Motor neuron

Muscle

Sensory neuron

Touch receptors in the skin are specialized to detect different kinds of touch. Some respond to temperature, others to pain and to pressure, and still others to the textures of what we touch.

Touch is as important to a baby's health and development as eating and sleeping.

Sensory Facts

- Being touched is important for normal human development. If babies are not cuddled a lot, they grow slowly, have trouble learning, and suffer physically. They also do not get emotionally close to other people.

- Many sensations result from combined signals to the brain. For example, no touch receptors sense wetness. Cold receptors and pressure receptors both send signals. The brain combines them to sense that the skin is wet.

- Goose bumps are the result of signals from cold receptors in the skin. The brain responds to cold with a message to the blood vessels and smooth muscles in the skin. The result is small bumps that help reduce heat loss.

- Itching is caused by an irritation to the skin. Irritants can be infections, allergies, insects, and fabrics. The irritant moves small touch receptors in the skin, which produces itching. Scratching can relieve the irritation. If scratching causes pain receptors to fire, that neutralizes the itching sensation.

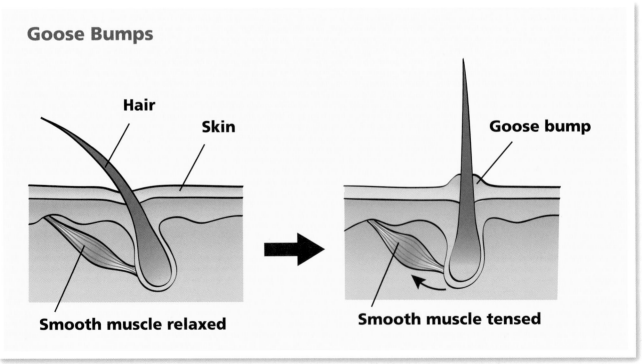

Goose Bumps

Hair

Skin

Goose bump

Smooth muscle relaxed

Smooth muscle tensed

Goose bumps and hair "standing on end" are automatic responses to cold or strong emotions like fear. Tiny muscles around hair follicles contract, bunching up the skin and pushing up the hairs.

- Say you have a sharp pain in a small area, such as a pinprick. Rubbing the area relieves pain. Rubbing activates pressure receptors in the skin. This decreases the intensity of the pain receptors' message to the brain.

- In the disease of leprosy, pain receptors are damaged. They do not send pain messages when parts of the body are hurt. As a result, people with leprosy do not know when they are harming their bodies. This lack of awareness can lead to a lot of damage, especially to hands and feet. Parts of the body that are damaged can become infected and inflamed. They can decay if untreated.

- The star-nosed mole has a great sense of touch. Its nose tentacles have six times more touch receptors than a human hand. As the mole tunnels in the dirt, its tentacles can touch 10 to 12 objects per second.

Think Questions

1. **Why don't you feel your watch, earrings, underwear, or socks after you put them on?**

2. **What do you think a web-building spider uses more to capture prey, its sight or sense of touch? Why?**

3. **What is the most useful thing you discovered about touch?**

The supersensitive nose of the star-nosed mole has a ring of 22 short, pink tentacles. They help it instantly identify prey by touch.

Hearing

Mechanoreceptors are involved in the **sense of hearing**. You hear by detecting vibrations.

The outer ear gathers and focuses sound waves. The waves vibrate the eardrum. The vibrations shake the tiny bones in the middle ear. The vibrations continue to the inner ear.

Mechanoreceptors in your inner ear are called hair cells. When hair cells receive vibrations, they send electric messages to the auditory nerve. The auditory nerve sends the message to the brain. The brain compares the messages to previous experiences. Comparing helps you determine what you heard. If the hair cells get damaged by loud sounds or disease, it affects your hearing.

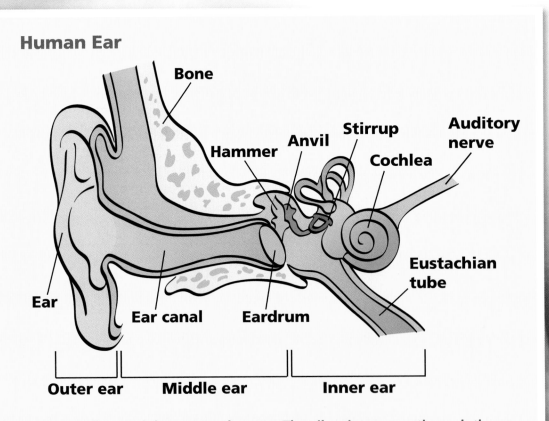

Human Ear

Bone

Hammer Anvil Stirrup

Cochlea

Auditory nerve

Eustachian tube

Ear

Ear canal Eardrum

Outer ear   Middle ear   Inner ear

Your ears collect and detect sound waves. The vibrations move through the ear canal to the inner ear, where they are converted to nerve impulses.

Hearing Loss

Hearing loss happens when some part of your ear is not working right. It can be caused by a head injury, a serious infection, or repeated exposure to loud sounds.

How good is your hearing? In 2014, scientists found that hearing loss among teens is on the rise. One out of six teens in the study had symptoms of hearing loss all or most of the time. Are you that one teen? You probably already know that listening to loud music is bad for your hearing. In the same 2014 study, scientists found that nine out of ten teens knew that they were doing things that can damage their hearing. Most teens also said that if their parents knew how loud their music was, they would tell them to turn it down. Have you been told to turn your music down? What is the loudest sound you've experienced?

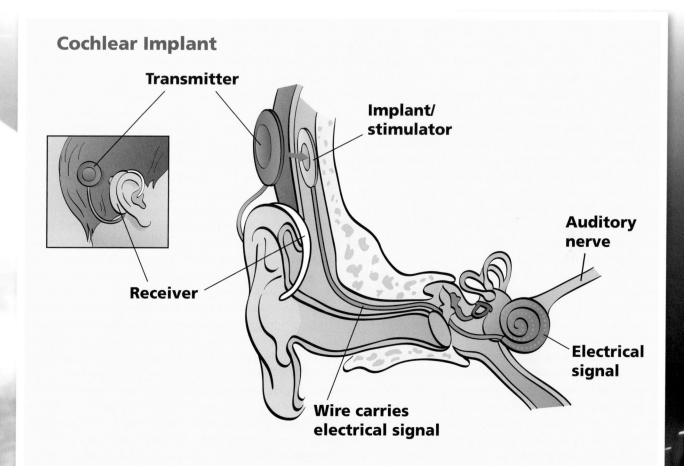

Cochlear Implant

Transmitter

Implant/ stimulator

Auditory nerve

Receiver

Electrical signal

Wire carries electrical signal

Unlike hearing aids, which amplify sounds, cochlear implants directly stimulate the auditory nerve. They can provide a limited sense of hearing to a person who is deaf or severely hard-of-hearing.

So what can you do to prevent hearing loss? Protect your (and your friends') ears from loud noises.

- Wear ear protection near loud machinery.
- If you can hear a friend's music over their earbuds or headphones, tell them to turn it down.
- If you play in a band, wear musicians' earplugs.
- If you hear ringing or buzzing in your ear, turn your music down.

Damaged hair cells can't be repaired. If you lose some, they are gone forever. Is there a cure for hearing loss? Hearing aids can amplify sounds. If hearing aids don't help, cochlear implants can help. The surgical implant takes over the job of the damaged hair cells. It turns sounds into electric signals that stimulate the hearing nerve.

Because headphones and earbuds deliver sound directly into the ear, music played loud and long can cause permanent hearing damage.

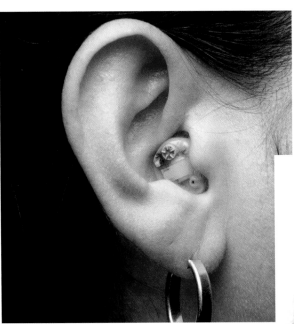

A hearing aid makes some sounds louder so they can be detected by damaged ears.

The sound processer and transmitter portion of a cochlear implant sits behind the ear.

Sensory Facts

- Hearing problems are the most common birth defect in babies.
- Hearing aids cannot restore normal hearing. The brain takes time to adapt to new sounds.
- When you hear your own voice as you talk, the sound travels through your jawbone and mouth as well as through the air in your ear. When you hear your voice on a recording, the sound travels only through air, so it sounds different.
- Sound volume is measured in decibels. A sound that is just barely audible is about 10 decibels. A jet airplane taking off is about 150 decibels. Every increase of 10 decibels represents 10 times as much sound. This means that a 20-decibel sound is 10 times as loud as a 10-decibel sound.

At 150 decibels, the sound of a jet airplane taking off is loud enough to rupture an eardrum.

Consider earplugs as protection against dangerously loud music and crowd noise at a concert.

- Some personal music players can produce sound levels between 120–130 decibels. Long exposures to sound levels above 80 decibels can permanently damage hair cells in the ear.
- Excessive noise can cause high blood pressure, a fast heart beat, emotional distress, learning difficulties, and disturbed sleep patterns. It can make you more likely to have accidents. In addition, high noise levels make workers less productive.
- Many animals, such as dogs and cats, can hear much higher frequencies than humans. Cats use this ability to locate their food. Mice make high-pitched sounds that cats can hear. That helps cats locate and catch the mice.

Think Questions

1. How are hearing and touch alike and how are they different?
2. Damaged hair cells can leak electric impulses into your brain. How could that affect your experience?
3. Which "Sensory Facts" information did you find most interesting? Why?

Smell and Taste

Have you noticed that when you have a cold, food doesn't taste as good? Smell and taste are separate senses, but they are closely related.

In the nose, odor molecules in the air stimulate chemoreceptors. Chemicals in food are detected by the chemoreceptors in your mouth and throat. The interactions of all the chemoreceptors in your nose, mouth, and throat sense flavor. When you have a cold, mucus coats the membranes in your nose so you can't smell, which means you can't taste as well either.

Odors are very powerful memory triggers. About 85 percent of people have some sort of childhood memory that is linked to a particular smell. One common memory is bread or cookies baking. Memories triggered by odors are often more emotionally intense than memories triggered by other senses. This intensity may be because the smell-processing part of the brain is close to the memory centers.

Take Note

Do you have a childhood memory linked to a particular smell? Ask a friend or relative if they have a childhood memory linked to a smell.

Smells, like that of fresh-baked cookies, can unlock memories of past experiences, much more so than sights or sounds can.

Sensory Facts on Smell

- If you want to identify a smell, take a deep breath. Air will swirl around inside your nasal cavity. It touches many different chemoreceptors. A deep breath improves the chances that the correct receptors will fire, sending information to the brain.
- Almost every organ in our body has chemoreceptors for smell. For example, the chemoreceptors in your skin can respond to a chemical in sandalwood. The chemical causes skin to heal faster.
- Some people can smell fear or disgust in other people's sweat. What they are actually smelling are chemicals released during times of intense emotion.

- Women generally have a better sense of smell than men.
- A decrease in sense of smell might be an early sign of Parkinson's or Alzheimer's disease.

Sweat itself is odorless, but chemicals released when excited or fearful may convey emotion.

Olfactory System

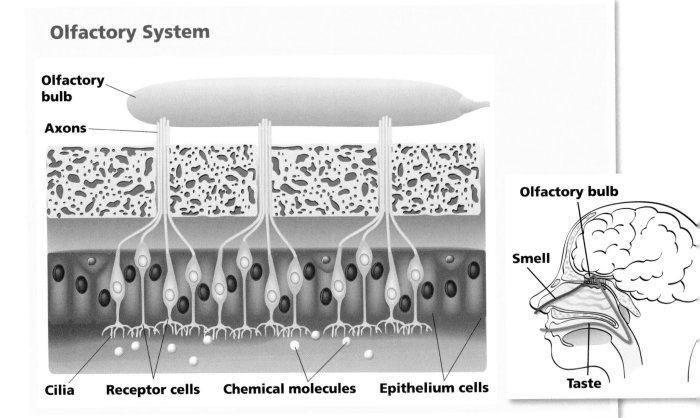

The human nose has 400 types of chemoreceptors. Each chemical molecule fits into a shape on a chemoreceptor and the receptor sends an electric message to a sensory neuron.

An elephant's keen sense of smell helps it detect water several kilometers away. Elephants also use scent to identify family and herd members.

- Smoking can seriously impair the sense of smell.
- Which animal do you think has the best sense of smell—a dog, an elephant, a cow, or a turtle? In a 2013 study, the Chinese soft-shelled turtle ranked 5th; cow, 4th; and dog, 9th. And the number one sniffer was the African elephant. Humans ranked a lowly 13th.
- Scientists think that the sense of smell is the most ancient sense. Even single-celled organisms can detect chemicals in their environment.

- Your sense of smell is much more sensitive than your sense of taste. Scientists have discovered that even a single molecule of some chemicals can trigger a smell response.

Even a one-celled amoeba has a sense of smell.

Sensory Facts on Taste

- It is a myth that different tastes are found on specific areas of the tongue. You can detect all tastes anywhere on the tongue.
- Taste buds last about 10 days; then new ones replace them.
- Babies have many more taste buds than adults. Their taste buds are all over the inside of their mouths, rather than concentrated on the tongue. As a result, babies and young children often prefer bland foods. As people get older, their sense of taste becomes less sensitive, so they can enjoy more flavorful or spicy foods.

Infants have about three times as many taste buds as adults, so eating is an intense experience.

Basic Tastes

Dark chocolate has a bitter taste.

Pretzels have a salty taste.

Pineapple has a sweet taste.

Meats have a savory taste called umami.

Lemons and limes have a sour taste.

The chemoreceptors of the taste buds on the tongue and other parts of the mouth can differentiate between basic tastes. Those tastes are bitter, salty, sweet, umami, and sour. There may be a receptor for fat as well.

Just by landing, a butterfly can tell if food is nearby. Receptors in its feet sense the dissolved sugars in flower nectar.

• Chili peppers cause pain receptors in the tongue and mouth to send messages to the brain. The messages stimulate the brain to release endorphins, natural pain suppressors. The endorphins produce feelings of pleasure, and possibly enhance the perception of other flavors in the food.

• Flies and butterflies have taste receptors in their feet. They can quickly taste whatever they land on.

Think Questions

1. Which "Sensory Facts" information did you find most interesting? Why? (Choose one for smell and one for taste.)

2. It is often stated that women are generally more sensitive to smell than men. Have you found this to be your experience? Poll family members and friends to get their opinions.

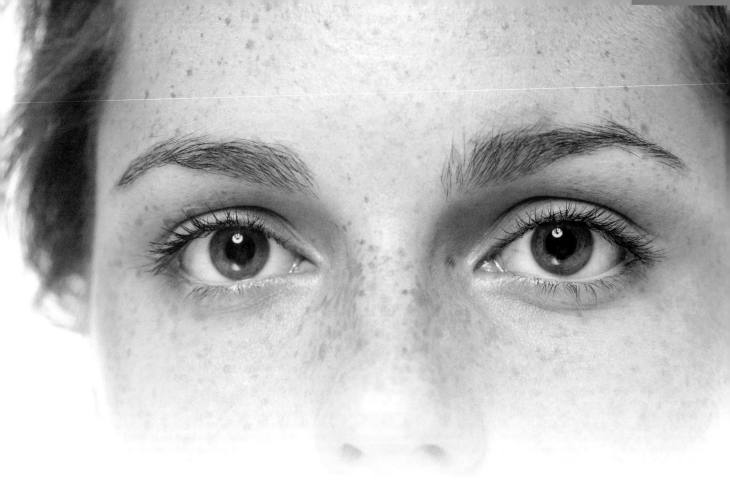

Sight

Look around you. What do you see? How do you see? Light bounces off objects in the environment and enters your eyes.

Light stimulates photoreceptors in the back of the eye. The receptor cells send messages to the brain through the optic nerve. And finally you see!

The structures of the eye have one important function: to focus an image on the retina. The retina is the curved screen at the back of the eye. This is where light energy changes into electric impulses that travel to the brain for processing. But how does that happen?

The Photoreceptors

The retina is actually an extension of the brain itself. When light reaches the retina, it passes through two layers of cells before hitting a dense layer of specialized light-sensitive cells, the photoreceptors. These cells transform light energy into electric impulses.

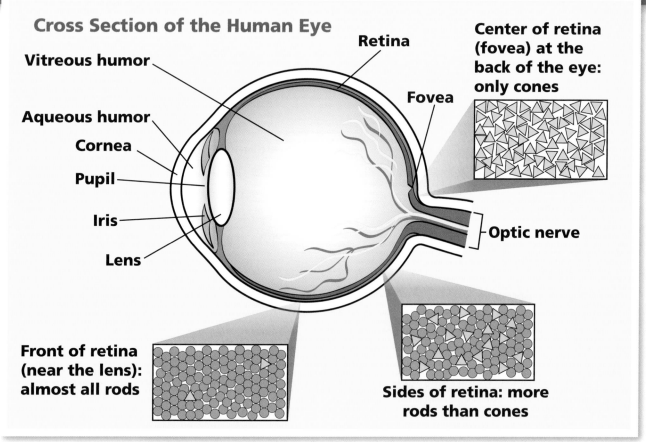

Cross Section of the Human Eye

Vitreous humor

Aqueous humor

Cornea

Pupil

Iris

Lens

Retina

Fovea

Center of retina (fovea) at the back of the eye: only cones

Optic nerve

Front of retina (near the lens): almost all rods

Sides of retina: more rods than cones

Each structure in the eye has a different function. Working together, the parts let us see the world around us . . . as long as there's light.

There are two types of photoreceptors, cones and rods. Cones detect color and detail. Cone cells are divided into three color sensitivities: red, blue, and green. Combined, the cones can detect about 100 different ranges of color. Your brain interprets these color ranges, so you "see" about 1 million colors. Cones are concentrated in the center of the retina at the back of the eye. They provide very detailed information to the brain.

Rods are sensitive to light and dark changes, movement, and shape. Rods allow vision in dimmer light. They help you see out of the side of your eyes.

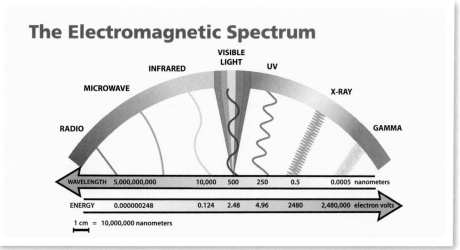

The Electromagnetic Spectrum

VISIBLE LIGHT

INFRARED

UV

MICROWAVE

X-RAY

RADIO

GAMMA

WAVELENGTH	5,000,000,000	10,000	500	250	0.5	0.0005 nanometers
ENERGY	0.000000248	0.124	2.48	4.96	2480	2,480,000 electron volts

1 cm = 10,000,000 nanometers

Only a small part of the electromagnetic spectrum is visible light.

An Inverted Image

When light enters the eye, the lens inverts it. This causes the image on the retina to be upside down. The brain flips the image back, so that you can see the world correctly. Imagine what it would be like to see the world upside down!

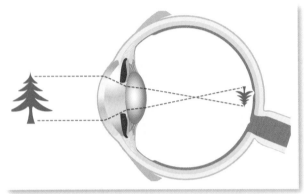

When the eye's lens focuses an image onto the retina, it also turns the image upside down.

Sensory Facts

- Humans have about 100 million rods and 5 million cones in the retina.
- A few women have a fourth type of cone in their eye, which means they are tetrachromatic (tetra = four; chrome = color). This fourth cone may be sensitive to orange light. This allows them to see more colors than most people. They can see more than 100 million different colors!
- Scientists found rods and cones in fossilized fish that may have lived in shallow waters 300 million years ago. This is the oldest record we have of rods and cones.

- In the same way that you are right-handed or left-handed, you are right-eyed or left-eyed. Here's how to find out which you are. Make a circle with the thumb and index finger of one hand. Hold your arm out in front of you. Look through the circle at a small object across the room. Close one eye and then the other. The eye that sees the object inside the circle is your dominant eye.

Take the simple circle test described in the text to see if you are right-eyed or left-eyed.

- When people look at something interesting, their pupils expand or dilate. A smart salesperson looks carefully into customers' eyes to see if they are interested.
- The iris gives the eye its color. All eye colors result from the same pigment, called melanin. Brown eyes have a lot of melanin. Blue eyes have very little melanin.

- Red-green vision problems affect about 1 in every 12 men and 1 in every 200 women. They are caused by a genetic mutation that affects the red and green cones. This condition is often called color blindness. True color blindness (not seeing any color at all) is very rare.

Take Note

Go to FOSSweb and engage with the resources in the "Vision Menu." Gather new information or clarify information in this reading.

Parts of the Human Eye

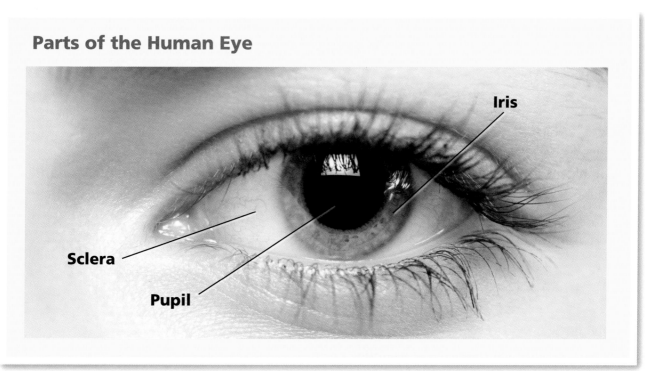

The visible eye has three parts: the sclera, or white of the eye; the dark pupil, or opening that lets light in; and the colored iris, which controls the amount of light entering through the pupil.

The compound eyes of insects bulge outward, allowing them to see in all directions at once. That's why it's hard to sneak up on a fly!

- The eyes of insects have hundreds or thousands of closely packed light-sensitive organs. The compound eye gathers light and focuses it. The brain of an insect puts together all these mosaic images to form a complete image.

- An eagle has about five times more rods and cones in its retina than a human. It has much sharper vision than we do. In addition, the area of greatest focus is a deep pit. This shape helps magnify an image. An eagle can spot a rabbit 5 kilometers (km) away.

Think Questions

1. **People can inherit traits that affect their cones. How does this affect their vision?**

2. **Which "Sensory Facts" information did you find most interesting? Why?**

Brain Messages

An ant is walking on your arm. The tickling gets your attention, and you raise your other arm to brush the ant away. How did you know where the ant was?

Your touch receptors for the tickle sensation were stimulated. They sent a signal to your brain, alerting you to a problem. Your brain decided how to respond. It sent messages to your arm, telling it how to brush away the ant.

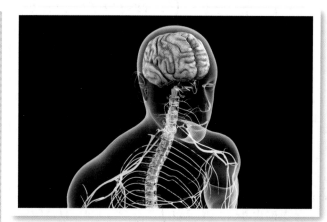

Your brain receives messages and decides how to respond.

Neurons at Work

The cells that make up your brain and all your nerves are called **neurons**. You have several hundreds of billions of neurons throughout your body and brain. They constantly send messages from one place to another.

Every neuron has three basic parts: a **cell body**, **dendrites**, and an **axon**. The cell body contains the nucleus of the cell. Extending from the cell body are numerous branches called dendrites (dendro = tree). Incoming electric messages enter the dendrites and travel to the cell body. Then they leave the cell body on a long, thin extension of the cell called the axon. The message passes from the axon to the dendrite(s) of one or more neurons. They carry the message on its way.

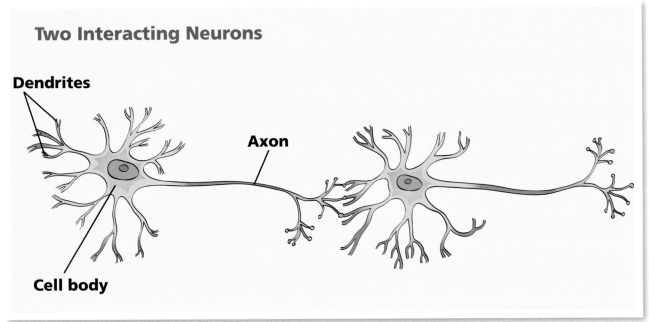

Two Interacting Neurons

Dendrites

Axon

Cell body

A message is picked up by dendrites, travels through the axon, and passes to the next neuron.

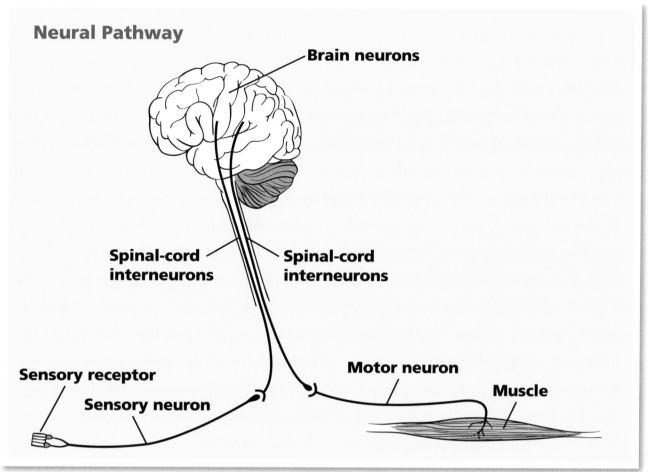

Neural Pathway

Brain neurons

Spinal-cord interneurons

Spinal-cord interneurons

Sensory receptor

Sensory neuron

Motor neuron

Muscle

The brain receives and sends the message that leads from the sensation to the response.

Neural Pathways

The sensory receptor cells for all the senses connect to the ends of sensory neurons. A signal travels from the sensory receptor to a sensory neuron, which connects to **interneurons** (*inter-* means "between") in the spinal cord. Where do the interneurons carry the message? To the brain.

The brain processes the message and decides what to do. If you want to move your arm to brush off an ant, the brain sends out messages to the muscles in your arms. The message travels down interneurons in the spinal cord to **motor neurons**. Where do the interneurons carry the message? Away from the brain, to motor neurons in the arm, and finally to the muscles. Your arm moves.

Take Note

Use the information in these paragraphs to label the structures on your *Neural Pathways* notebook sheet.

Nerve impulses travel through massive networks of neurons. These impulses control virtually everything in the body. The brain coordinates it all.

Communication

Sensory neurons, motor neurons, and interneurons are all like wires carrying an electric signal. Sensory neurons carry information *to* the brain, and motor neurons carry messages *from* the brain. Interneurons are between the other neurons and the brain. The motor neurons carry instructions to the muscles. Your arm muscles respond by contracting. Stretch receptors in the muscle then give the brain information on how much the muscles are stretched or contracted.

This kind of communication is happening constantly between your brain and the rest of your body. The flow of electric signals along neurons allows you to sense the environment, make decisions, move, breathe, and so on.

One cubic millimeter of brain tissue contains 1 billion neural connections.

Reaction Time

Sending messages takes time. The longer the pathway, the longer it takes for a stimulus to produce a response. The interval between stimulus and response is sometimes called **reaction time**. You may have noticed this delay if you have ever stubbed your toe. You can see it being stubbed and hear the sound before you feel the pain! The pathway from your eyes and ears to your brain is much shorter than the pathway from your toes to your brain. So the sensory neurons in your eyes and ears get their messages to the brain before the sensory neurons in your toes can.

Neurotransmission: The Body's Amazing Network

Inside your body is a living communication system. It is more complex than the most advanced computer network.

Your brain and nerves send and receive billions of electric and chemical messages every minute. They connect everything, from the depths of the brain to the remote toes. Let's take a closer look.

From Neuron to Neuron

Remember that neurons are cells that carry the messages within the nervous system. The messages they carry are electric impulses.

Neurons pass their messages from one to another. But they do not touch each other. Instead, a tiny gap called a **synapse** separates the axon of one neuron and the dendrite of the next neuron.

Once an electric impulse reaches the end of an axon, it stimulates the release of chemicals into the gap. These **neurotransmitters** flow across the synapse. They fit into surface receptors of the receiving dendrite.

The neurotransmitters trigger the continuation of the electric message in the receiving dendrite. The electric impulse races down the axon. The process repeats, relaying the message on its way.

Take Note

Trace the pathway of one electric impulse from the axon of one neuron to the dendrite of the next. State what is happening at each step.

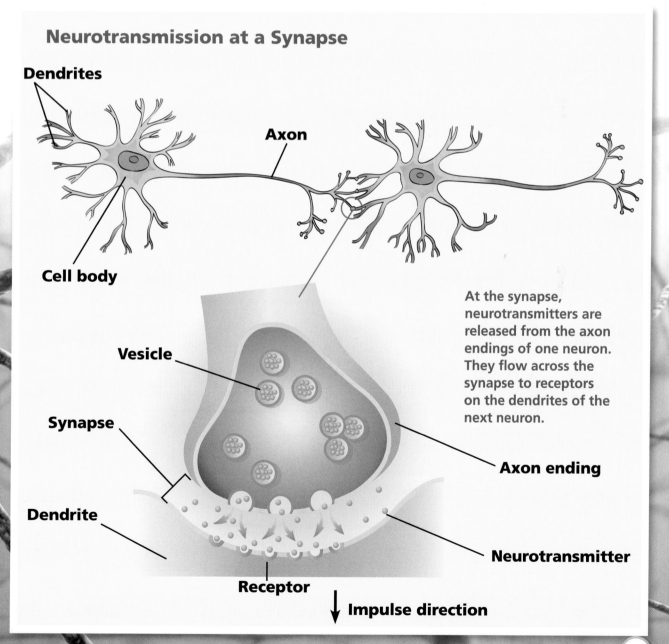

Neurotransmission at a Synapse

Dendrites

Axon

Cell body

Vesicle

Synapse

Dendrite

At the synapse, neurotransmitters are released from the axon endings of one neuron. They flow across the synapse to receptors on the dendrites of the next neuron.

Axon ending

Neurotransmitter

Receptor

↓ **Impulse direction**

The Communication Network

The typical human brain has 100 billion neurons, and 1,000 trillion neural connections. This incredible number of pathways is responsible for everything a person learns, remembers, says, sees, and does. In fact, it is responsible for everything that makes a person human. A baby's brain is building those pathways even before it is born!

During the first few years of life, the brain continues to develop systems of interactions and communication among all its neurons. Every time a baby looks at something, information enters his or her brain and forms a pathway between neurons. With repetition, that pathway gets stronger and those neurons thrive. Neurons that are not used do not become connected in strong permanent pathways, and they eventually die.

An electroencephalogram (EEG) is a test that measures electrical activity in the brain. Even in infants, an EEG can detect abnormal brain waves and help doctors evaluate brain disorders.

Every neuron in the brain is connected to as many as 15,000 other neurons, forming an incredibly complex network of neural pathways. The foundation for the network is laid in early childhood.

What does the formation of neural pathways have to do with memory formation?

Why do people say "use it or lose it"?

Many mysteries remain in our understanding of neural networks. For example, scientists are studying the birth and death of neurons. This work may lead to new treatments for brain diseases and disorders. Studies are showing that a teen's intellectual power is a match for an adult's. Your capacity to learn will never be greater than now. Build those neural pathways and take advantage of your brain power now!

Think Questions

1. **The relay we did in class is a model for this process. Refer to the illustration on page 85. Compare the relay to the actual transmission of a message. What does each part of the relay represent?**
2. **Communication in the nervous system is called electrochemical. Why is this so?**

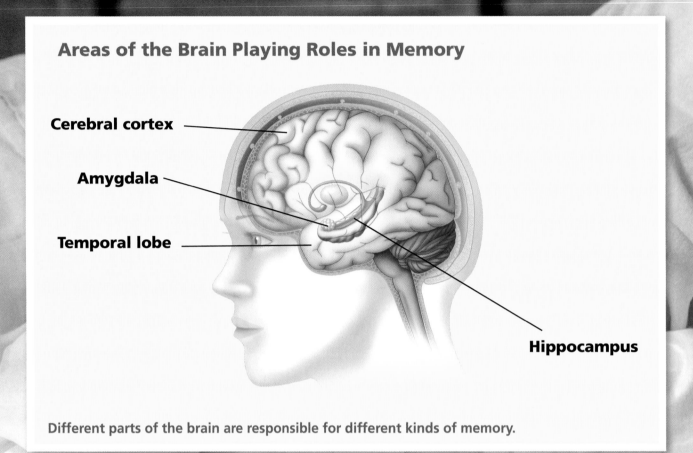

Areas of the Brain Playing Roles in Memory

Cerebral cortex

Amygdala

Temporal lobe

Hippocampus

Different parts of the brain are responsible for different kinds of memory.

Memory and Your Brain

When you form a memory, the brain physically changes. Several key parts of the brain help form and store memories.

Most types of memory appear to be stored in the **cerebral cortex**, the outer layer of the cerebrum. The **hippocampus** and the **temporal lobe** help you learn general facts or remember events. And the **amygdala** is involved in emotional memory.

Typically, within the first minutes or hours of learning, neurons change. They form connections with other neurons. The connections allow communication between the cells. The more you practice or study, the stronger these networks get.

The neural connections that form our memories are strengthened during sleep.

What Affects How Memories Form?

Four things affect how memories form. They are paying attention, getting enough sleep, emotions, and the number of senses.

Paying Attention

"Why don't you remember what I told you?"

"I wasn't paying attention."

That's the problem. Doing many things at once divides our attention. Much research has focused on media multitasking. It looks at students who use multiple kinds of technology at the same time. Guess what the studies are finding? These students make many mistakes in memory tests. They are too distracted to remember what happened. So cut out distractions when you are trying to learn new things.

Multitasking is not as efficient as it sounds. When your brain is constantly switching focus and responding to interruptions, you'll find it hard to concentrate when you need to.

Getting Enough Sleep

A 2014 study investigated how sleep affects mouse neurons. When mice learn new things, spines appear on the dendrites of neurons they use. These spines can connect to other neurons. The learning or memory is stored. When mice sleep after the new tasks, those neurons grow more than if they don't sleep.

This study concludes that sleep contributes to storage of new memories. Your memory will work better if you get enough sleep. And that means you will do better on tomorrow's test if you get a good night's sleep.

In humans and other animals, sleep has a profound, positive effect on learning and memory.

Emotions

The amygdala helps us recognize emotionally significant events. You may not often remember what you had for dinner a week ago. But say you heard some really bad news during dinner. You may remember the details of that event, including what you were eating. Positive emotions can also affect how well you remember an event, but negative emotions are even more powerful.

Take Note

Think back to when your teacher asked you to remember your oldest memory. What emotions were associated with the memory you recalled?

More Senses Help Learning

Remember the memory tasks you tried? Which ones were easiest for you? For most people, the more senses that are used at the same time, the more they can remember. By writing things down or saying them out loud, you can learn better than by simply reading. Different kinds of neural networks form. If you can't remember something you saw, maybe you can remember what you heard or wrote.

If too many senses are stimulated at the same time, however, it can be confusing. Your brain cannot process too much information at once.

Our brains are incredibly complex and always changing. As we learn and remember experiences, the brain builds networks of neurons that maintain that learning and memory. Those neural networks help make us who we are.

Think Questions

1. **How do you best study for a test or learn new information?**
2. **How can you improve your memory? What strategies have you heard of besides the ones in this article?**

Taking notes can help you remember what you see and hear in class.

Images and Data

Images and Data   Table of Contents

Diabetes Affects Human Organ Systems

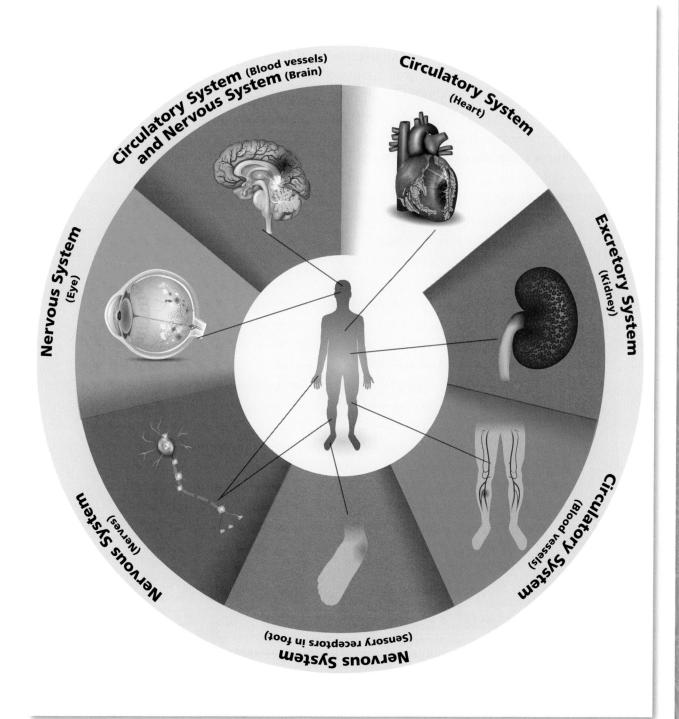

Disease Information

Hantavirus pulmonary **syndrome** is a severe respiratory disease in humans caused by infection with a hantavirus. Early symptoms may be mild, and the disease is easily confused with the flu. Rodents carry the virus, most commonly deer mice. People become infected when they breathe in air that has been exposed to mouse droppings or urine. Working wherever deer mice live increases your chance of being infected with hantavirus. This includes cleaning unused buildings like garages or cabins. Camping where deer mice are common also increases the risk of infection.

Hantavirus pulmonary syndrome occurs mainly in North, South, and Central America. Related viruses cause disease in other parts of the world, particularly Asia.

Deer mice infected with the hantavirus (inset) appear healthy and normal, so it is wise to avoid all contact with wild rodents, their droppings, urine, and nests.

Symptoms

Hantavirus pulmonary syndrome can be fatal. Symptoms usually develop 1 to 5 weeks after exposure to the virus in rodent droppings or urine.

Early symptoms
- Fatigue
- Fever and chills
- Muscle aches, especially in large muscles such as the back, thighs, and shoulders
- Headaches and dizziness
- Nausea, vomiting, diarrhea, and abdominal pain

Later symptoms
- Coughing and shortness of breath
- Tightness in chest
- Low blood pressure

Treatment

There is no specific treatment, vaccine, or cure for hantavirus. Patients who see a doctor early have a higher chance of survival. Though there is no cure, doctors can treat the symptoms. For example, patients are often given oxygen to help them breathe.

Did You Know?

The discovery of hantaviruses is relatively recent. Hantavirus pulmonary syndrome was first noticed in soldiers during the Korean War in the 1950s. The virus has probably been around for a lot longer.

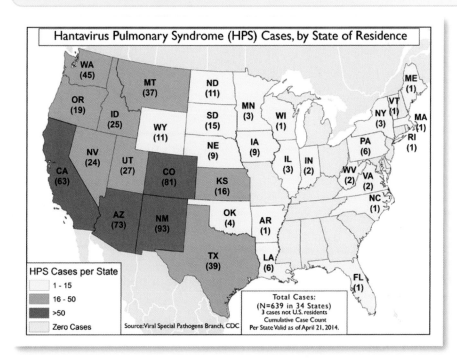

Hantavirus Pulmonary Syndrome (HPS) Cases, by State of Residence

WA (45) • MT (37) • ND (11) • ME (1) • OR (19) • ID (25) • WY (11) • SD (15) • MN (3) • WI (1) • NY (3) • VT (1) • MA (1) • RI (1) • NV (24) • UT (27) • CO (81) • NE (9) • IA (9) • IL (3) • IN (2) • PA (6) • WV (2) • VA (2) • CA (63) • KS (16) • NC (1) • AZ (73) • NM (93) • OK (4) • AR (1) • TX (39) • LA (6) • FL (1)

HPS Cases per State
- 1 - 15
- 16 - 50
- >50
- Zero Cases

Source: Viral Special Pathogens Branch, CDC

Total Cases:
(N=639 in 34 States)
3 cases not U.S. residents
Cumulative Case Count
Per State Valid as of April 21, 2014.

Lupus

Lupus is a chronic disease. It lasts for a lifetime and keeps coming back. It can affect any organ in the body. Lupus occurs when something goes wrong with the immune system. The immune system usually helps the body fight off harmful bacteria and viruses. But lupus causes the immune system to attack healthy tissue. The attacks cause pain and inflammation (parts of the body become red, swollen, and hot). A family history of lupus increases a person's risk of developing lupus.

Immune System

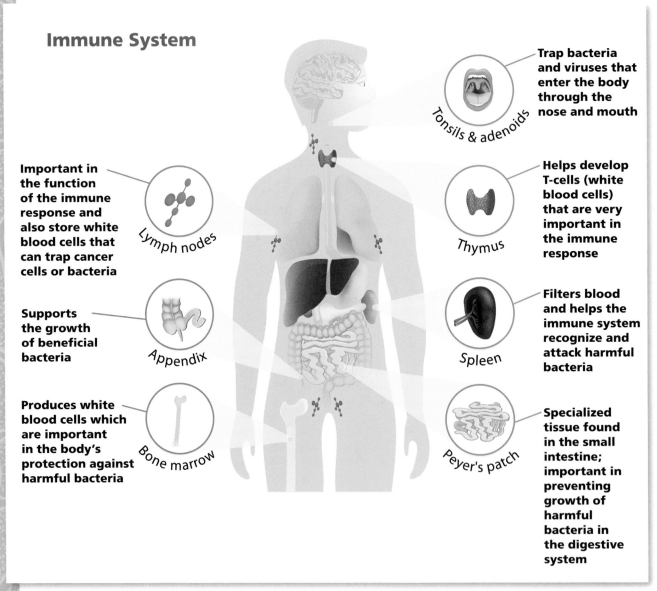

Tonsils & adenoids — Trap bacteria and viruses that enter the body through the nose and mouth

Thymus — Helps develop T-cells (white blood cells) that are very important in the immune response

Spleen — Filters blood and helps the immune system recognize and attack harmful bacteria

Peyer's patch — Specialized tissue found in the small intestine; important in preventing growth of harmful bacteria in the digestive system

Lymph nodes — Important in the function of the immune response and also store white blood cells that can trap cancer cells or bacteria

Appendix — Supports the growth of beneficial bacteria

Bone marrow — Produces white blood cells which are important in the body's protection against harmful bacteria

The different parts of the immune system work together to keep your body healthy.

Symptoms

Lupus is difficult to diagnose because its symptoms are often like those of other diseases. It occurs more often in women than in men. Symptoms can be triggered by sunlight, certain medications, and infections.

Common symptoms
- Fatigue and fever
- Pain, especially in the joints and chest
- Swelling in joints
- Shortness of breath
- Headaches, confusion, and memory loss
- Dry eyes
- Rash on the cheeks and nose, often described as a "butterfly" rash (not seen in all patients)

Treatment

Lupus cannot be cured. Treatment helps minimize damage to organs, so it is important to be diagnosed early. Medication that reduces swelling can be helpful. Avoiding sunlight and sugary foods can improve some symptoms.

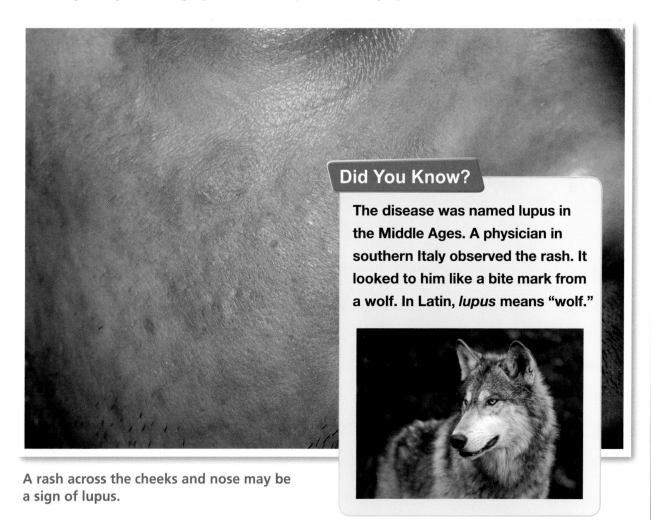

Did You Know?

The disease was named lupus in the Middle Ages. A physician in southern Italy observed the rash. It looked to him like a bite mark from a wolf. In Latin, *lupus* means "wolf."

A rash across the cheeks and nose may be a sign of lupus.

Lyme Disease

Lyme disease is caused by bacteria carried by blacklegged (deer) ticks. You can pick up these ticks when you walk through woods or grassy areas. After an infected tick bites a person, the bacteria can enter the wound. If left untreated, the bacterial infection can spread to the joints, the heart, and the nervous system.

Lyme disease is rapidly increasing in the United States. Cases of Lyme disease have been reported in all states and many Canadian provinces. It is most common in the northeastern and midwestern United States.

Reported Cases of Lyme Disease—United States, 2015

Though Lyme disease cases have been reported in nearly every state, cases are reported based on the county of residence, not necessarily the county of infection.

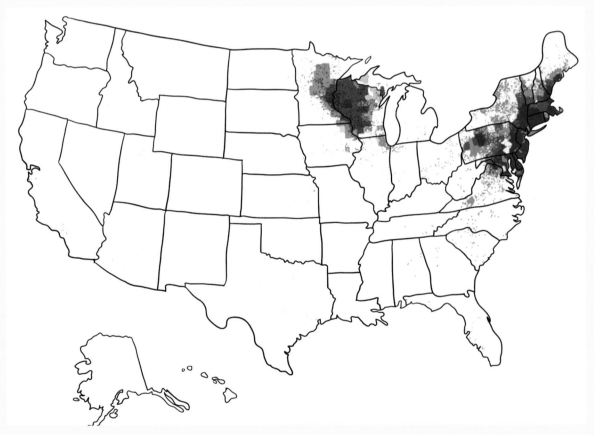

One dot is placed randomly within the county of residence for each confirmed case.

Symptoms

Early symptoms
- Rash
- Fever
- Chills
- Fatigue
- Body aches
- Headaches

Later symptoms
- Expanding bull's-eye rash
- Severe joint pain and swelling
- Weakness in limbs
- Heart problems
- Eye inflammation
- Paralysis of facial muscles

Did You Know?

Lyme disease was first recognized in the United States in the 1970s. A group of people in Lyme, Connecticut, had some puzzling symptoms. It was gradually determined that the affected people had all been bitten by blacklegged ticks. The bacterium that causes Lyme disease (*Borrelia burgdorferi*) was classified in 1981.

Treatment

If Lyme disease is diagnosed early, antibiotic treatment is usually effective. But even after treatment, a small number of people still experience symptoms. It is thought that the initial infection might trigger the immune system to attack healthy tissue. This may cause the continuing symptoms.

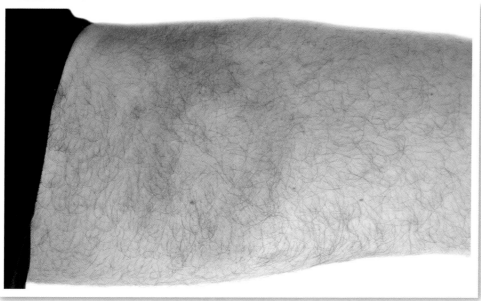

The characteristic bull's-eye rash of Lyme disease is not painful and does not itch.

 Where to find more information about these three diseases
Center for Disease Control and Prevention
Mayo Clinic

Brain Map of Sensory Activity

The image shows where different sense information is processed. These regions are on the outer surface of the brain, the cerebral cortex.

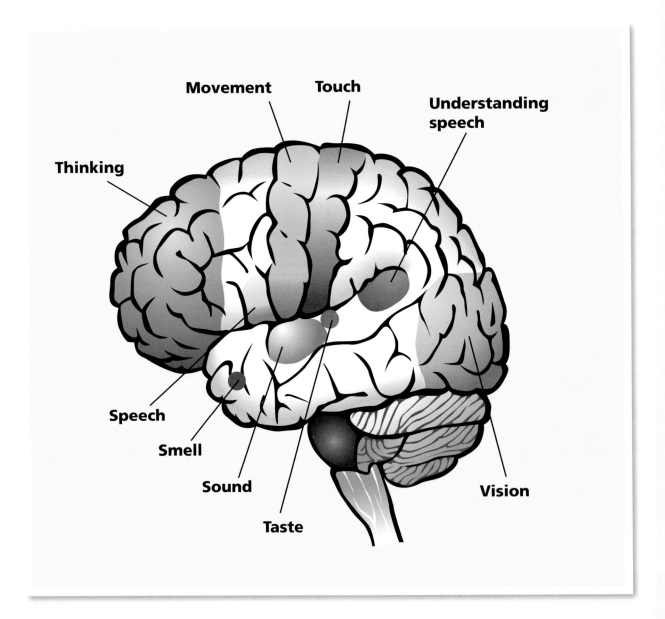

Science Safety Rules

1. Always follow the safety procedures outlined by your teacher. Follow directions, and ask questions if you're unsure of what to do.

2. Never put any material in your mouth. Do not taste any material or chemical unless your teacher specifically tells you to do so.

3. Do not smell any unknown material. If your teacher asks you to smell a material, wave a hand over it to bring the scent toward your nose.

4. Avoid touching your face, mouth, ears, eyes, or nose while working with chemicals, plants, or animals. Tell your teacher if you have any allergies.

5. Always wash your hands with soap and warm water immediately after using chemicals (including common chemicals, such as salt and dyes) and handling natural materials or organisms.

6 Do not mix unknown chemicals just to see what might happen.

7 Always wear safety goggles when working with liquids, chemicals, and sharp or pointed tools. Tell your teacher if you wear contact lenses.

8 Clean up spills immediately. Report all spills, accidents, and injuries to your teacher.

9 Treat animals with respect, caution, and consideration.

10 Never use the mirror of a microscope to reflect direct sunlight. The bright light can cause permanent eye damage.

Glossary

abnormal different from what is usual

aerobic cellular respiration the process by which a cell releases energy using chemical reactions that require oxygen

alveolus (plural: alveoli) tiny air sac in the lungs, surrounded by capillaries, where oxygen and carbon dioxide are exchanged

amygdala (ah•MIG•da•la) a region of the brain involved in emotional memory, such as fear

artery a muscular blood vessel that carries blood from the heart to the body

autonomic nervous system the system that controls the functions of internal organs and is not consciously directed

axon a thin and usually long extension of the neuron that carries impulses away from the cell body

bone marrow a spongy material found in the center of most bones

capillary the smallest blood vessel where gases, nutrients, and wastes are exchanged between blood and cells.

cardiac muscle the muscle tissue of the heart

cartilage a connective tissue in joints between bones

cell the basic unit of life

cell body the operational center of the neuron, which contains the nucleus of the cell

central nervous system part of the nervous system made up of the brain and spinal cord

cerebral cortex the outer layer of the cerebrum where millions of neurons make sense out of the signals that come into the brain

chemoreceptor a sensory cell that responds to a chemical stimulus

circulatory system the system of blood vessels and organs that transports blood to the cells in the body

cone a photoreceptor in the eye that distinguishes color and detects fine details in bright light

dendrite a branch that extends from a neuron's cell body and receives incoming information

digestive system the organs and structures that process food in the body

electromagnetic light input that photoreceptors respond to

endocrine system the system that makes, stores, and releases hormones

enzyme a protein that regulates chemical reactions

epiglottis a flap that directs food down the esophagus and away from the trachea

excretory system the organs and structures, responsible for the elimination of waste from the body.

fatigue an extreme lack of energy

gland a specialized group of cells that manufactures and releases hormones

glucose a simple sugar that is an important energy source

heart a muscular organ that pumps blood

hippocampus a region of the brain related to memory

homeostasis the constant regulation of internal conditions in the body

hormone a substance produced in the body that helps to control the way a cell or organ works

interneuron a neuron that connects sensory and motor neurons

joint where bones meet; allows the body to move in different ways

mechanoreceptor a sensory cell that responds to mechanical stimuli such as pressure or sound waves

metabolism all of the chemical reactions that take place in the human body

motor neuron a nerve cell that sends information from the brain or spinal cord to a muscle or gland

muscular system the system that gives shape to the body and allows it to move; made up of skeletal muscles, smooth muscles, and cardiac muscle.

nervous system the system that controls all the activity inside the body, and monitors and responds to the outside environment

neuron a nerve cell that transmits electrical impulses

neurotransmitter a chemical that is released when an electric impulse reaches the end of a neuron's axon; transmits the signal to the dendrites of the next neuron

osteoblast a bone cell responsible for making new bone tissue for growth or repair

pain a message created by a sensory receptor in response to potentially harmful stimuli

peripheral nervous system part of the nervous system; made up of all the nerves outside of the brain and spinal cord

peristalsis the smooth muscle activity that pushes food from the esophagus to the stomach

photoreceptor a sensory cell that responds to visible light in the electromagnetic spectrum

photosynthesis the process by which organisms that have chlorophyll use light energy, carbon dioxide, and water to make sugar

plasma the fluid part of blood containing red blood cells, white blood cells, and platelets

platelet a type of blood cell that is important for blood clotting

pressure mechanical stimulation of the skin

reaction time the time it takes for a stimulus to produce a response

receptor a cell that collects information to send to the brain for interpretation

red blood cell a blood cell that contains hemoglobin, which transports oxygen and carbon dioxide

respiratory system the organs and structures that transport oxygen to the red blood cells and get rid of carbon dioxide.

rod a photoreceptor in the eye that becomes active in dim light

saliva fluid produced in the mouth that aids digestion

sense of hearing the perception of vibration experienced through the outer ear and eardrum

sense of sight the perception of light experienced through the eyes

sense of smell the perception of chemicals experienced through the nose

sense of taste the perception of chemicals experienced through taste buds in the mouth

sense of touch the perception of pressure, temperature, or pain experienced through the skin

sensory neuron a nerve cell that sends information from sense receptors to the brain

sensory receptor a nerve ending that responds to a stimulus in the environment

skeletal muscle a single organ of muscle tissue, blood vessels, tendons, and nerves that produces contractions resulting in movement

skeletal system a system of bones that provides structure, assists movement, and protects the main organs of the nervous system

smooth muscle involuntary muscle tissue found inside organs, such as the stomach, intestines, and blood vessels

spinal cord the part of the nervous system that carries information between the brain and other parts of the body

stimulus (plural: stimuli) anything that causes an action or response

symptom an indicator that something is wrong in the human body

synapse a tiny gap between the axon of one neuron and the dendrite of another neuron; the point at which a nerve impulse passes from one neuron to another by way of a chemical neurotransmitter

syndrome a collection of symptoms produced by a disease

temporal lobe one of the four main lobes of the cerebral cortex where memories are stored and language processed

tendon ropelike tissue that connects muscle to bone

vein a blood vessel that carries blood from the body to the heart

white blood cell a blood cell that defends the body against disease

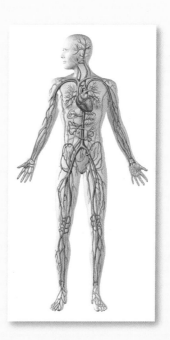

Index

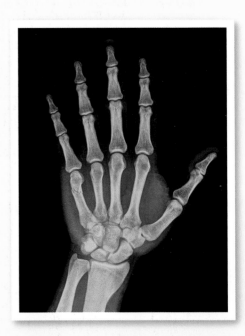